GIS TECHNOLOGY APPLICATIONS IN

ENVIRONMENTAL AND EARTH SCIENCES

GIS TECHNOLOGY APPLICATIONS IN

ENVIRONMENTAL AND EARTH SCIENCES

BAI TIAN

CRC Press
Taylor & Francis Group
Boca Raton London New York

CRC Press is an imprint of the
Taylor & Francis Group, an **informa** business

CRC Press
Taylor & Francis Group
6000 Broken Sound Parkway NW, Suite 300
Boca Raton, FL 33487-2742

First issued in paperback 2019

ISBN-13: 978-1-4987-7604-2 (hbk)
ISBN-13: 978-0-367-88959-3 (pbk)

Library of Congress Cataloging-in-Publication Data

Names: Tian, Bai, author.
Title: GIS technology applications in environmental and earth sciences / Bai Tian.
Other titles: Geographic information systems echnology applications in environmental and earth sciences
Description: Boca Raton, FL : Taylor & Francis, 2017. | Includes bibliographical references and index.
Identifiers: LCCN 2016013748 | ISBN 9781498776042 (alk. paper)
Subjects: LCSH: Environmental sciences--Remote sensing. | Environmental sciences--Remote sensing--Case studies. | Earth sciences--Remote sensing. | Earth sciences--Remote sensing--Case studies. | Geographic information systems. | Geographic information systems--Case studies.
Classification: LCC GE45.R44 T53 2017 | DDC 550.285--dc23
LC record available at https://lccn.loc.gov/2016013748

Visit the Taylor & Francis Web site at
http://www.taylorandfrancis.com

and the CRC Press Web site at
http://www.crcpress.com

This book is dedicated to my parents, three sisters and two brothers,

my wife (Yi Li), my daughter (Stephanie Tian), and my son

(George Tian) for their inspiration, encouragement, and help.

Being the first in my family to attend college in difficult times, I could not have made it without the encouragement, support, and sacrifices of my parents and siblings. Although my parents were too poor to go to school and virtually illiterate, they understood the value of education and helped me to achieve the education they had dreamed of. My elder brother and sister quitted school just to help the family to get by. Their spirit of hard work and selflessness has inspired me in every aspect of my life. To them, I owe a lot.

My wife, daughter, and son helped me greatly in the writing of this book. They supplied me with new ideas and perspectives, encouraged me when I was tired of writing, and did most of the house chores so that I could focus more time and energy on this book. In particular, my daughter Stephanie, an outstanding artist and scholar, designed this book cover and helped me in the editing of the book chapters. Without their encouragement and support, I might never have started or finished this project. To them goes my big appreciation. Cheers!

Contents

Preface

Since the late 1990s, Geographic Information System (GIS) technology has become increasingly popular and is now widely used in environmental and earth sciences. It is a huge and fast-growing industry and market worldwide, with enormous demand for GIS knowledge, experience, data, and software products. Professors need to know what the real-world needs are for GIS so that they can design better teaching and lab materials for their students to prepare them for their future careers. Students want to see what real-world projects look like, and what skills and knowledge they have to learn in order to make themselves more competitive in a tough job market. Environmental scientists, project managers, legislators, attorneys, activists, and the public are curious about what GIS technology is and how it can help with their projects or cases. GIS professionals are interested in learning the basics of environmental and earth sciences in order to better apply GIS technology in these and other related disciplines. However, there is a shortage of high-quality books that cover both GIS technology and its applications in environmental and earth sciences because not many people have the in-depth knowledge, experience, and data in multiple academic disciplines required to write them. This book was written to bridge such knowledge gaps and to meet the demands of a wide range of readers with different levels of GIS knowledge and skill. It intends to offer readers big pictures, new ideas, and innovative approaches.

The author is an interdisciplinary expert on GIS, and environmental and earth sciences, involved in hundreds of projects with over a thousand project sites across the United States, its territories, and other countries worldwide, as a student, geologist, environmental scientist, and GIS manager over the past 30 years. Among them are a variety of high-profile projects with national and international significance, such as the Libya chemical weapons disposal project, which attracted worldwide attention and was reported by the *New York Times* on February 2, 2014, entitled "Libya's Cache of Toxic Arms All Destroyed."

The author is also one of the pioneers who introduced GIS technology into environmental and earth sciences and contributed to its development. Since the early 1990s, he has presented and written numerous papers based on the large number of projects he has worked on and managed in the past 30 years. In late 2013, the author was invited to a United Nations GIS conference in New York City, which was attended by leaders and experts representing UN agencies, the World Bank, national and regional governments, academia, and related industries. During the conference, the author discussed with the leaders and experts his success stories of using GIS technology to help identify and solve environmental problems in a variety of projects worldwide. They were very impressed with his interdisciplinary and innovative GIS approaches and encouraged him to write a book to share his GIS knowledge and experience with the worldwide community. Inspired by these world-renowned leaders and experts, the author wrote this book to systematically introduce GIS technology and to demonstrate how GIS can be used to understand problems and make better decisions to solve them.

This book was written based on in-depth knowledge, experience, data, and lessons learned from a wide range of real-world projects over the past 30 years. It was also written from the unique perspectives of a GIS learner/student, professional, and manager, and a geologist, as well as an environmental scientist. It covers multiple scientific disciplines and contains many case studies extracted from a large number of real-world projects.

The book starts with an easy-to-understand overview of GIS technology for readers of all levels, such as professors, students, GIS professionals, environmental scientists, geologists, hydrologists, geophysicists, civil engineers, project managers, leaders, activists, attorneys, and the general public, to have an understanding of what GIS technology is; how GIS can help with their projects, studies, or concerns; what free or commercial GIS software products and data sources are available worldwide; what spatial data and metadata standards are available to follow; and so on. Throughout the book, the author explains (using case studies, computer programs, maps, graphics, and 3D models) how GIS and other related technologies can be used to automate mapping processes; collect, process, edit, store, manage, and share datasets; and statistically analyze data, model, and visualize large datasets to understand patterns, trends, and relationships in order to make educated decisions. This book is an excellent comprehensive resource for readers interested in learning about and using GIS and related technologies in environmental and earth sciences, geography education, natural/cultural resources conservation, land management, mining, oil/gas exploration, civil engineering, planning, disaster assessments and responses, humanitarian aid, utilities, and many more fields.

Chapter 1 is a complete, but easy to understand, overview of GIS technology, including its definition and history, data types, coordinate systems and projections, spatial data and metadata standards, free and commercially available data sources and GIS software products worldwide. Chapter 2 explains how to use GIS and GPS technologies to collect, edit, process, store, and manage data more efficiently while ensuring data quality, integrity, and security. Chapter 3 focuses on data processing and statistical analysis to extract useful information and reveal relationships, patterns, and trends to better understand problems and make educated decisions. Chapter 4 discusses how to use GIS technology in geologic mapping and data analysis, including converting paper maps into digital GIS maps and data layers through digitizing and scanning techniques, processing and plotting geologic measurements onto maps automatically, constructing 3D geologic models/maps and visualizing them with Virtual Reality (VR) technology, and so on. Chapter 5 demonstrates how to use GIS and other related software applications in 3D modeling and visualization of geologic and environmental data to better understand project site conditions and design the most appropriate investigation, mitigation, and/or remedial strategies. Chapter 6 introduces how to share GIS data and maps through the Internet, including a comprehensive review of a variety of free and commercially available data sources (i.e., websites and FTP sites) worldwide, serving project data and maps over the Internet, and so on.

Professors can refer to some chapters or sections of this book in designing and preparing their teaching and lab materials in mapping, data analysis, 2D and 3D modeling, visualization, and so on. Students can read the whole book to learn how to use GIS technology to understand and solve real-world problems. Project managers and other decision-makers can read certain chapters just to get a big picture of what GIS technology is and how it can help with their projects. Environmental attorneys, activists, and other interested and concerned citizens can learn some GIS basics and how to use free GIS software to collect and analyze data to support their cases and tell stories. The author hopes that readers will learn something new and useful from this book and be inspired both technically and spiritually.

Acknowledgments

The author would like to thank the following organizations:

- The United Nations Institute for Training and Research (UNITAR)
- The United Nations Operational Satellite Application Program (UNOSAT)
- The European Environment Agency (EEA)
- U.S. Army Corps of Engineers (USACE)
- U.S. Navy Engineering Field Activity Northeast
- U.S. Geological Survey (USGS)
- U.S. National Park Service (USNPS)
- U.S. Environmental Protection Agency (USEPA)
- Parsons Corporation
- Department of Geological and Atmospheric Sciences of Iowa State University
- Iowa State University GIS Lab
- EA Engineering, Science, and Technology, Inc.
- East China Institute of Technology
- Chinese Academy of Geosciences
- Xi'an Institute of Geology and Mineral Resources
- Ministry of Science and Technology of PRC
- Japan International Cooperation Agency (JICA)

Author

The author was originally trained as a geoscientist in the early 1980s. He started his career as a geologist, studying the plate tectonics of the Himalayas, Tibetan Plateau, and other related regions. With the Chinese economic boom of the late 1980s came serious air, soil, and water pollution from mining, manufacturing, power plants, farming, transportation, burning of coal and fossil oil, and overcapacity of the aging sewer systems; and lagging regulations, monitoring, and enforcement, and so on. Environmental investigations and remediation became a higher priority for the country. Therefore, the author began work on environmental projects, such as air-quality monitoring of the Copper Valley Gold Mine, evaluating the environmental impact of the Wei River Power Plant and designing remedial alternatives, investigating a polluted river by a paper mill, and other projects. For each geologic or environmental project, a large quantity of data was collected, processed, analyzed, and compiled into maps, graphics, and summary tables to help scientists, stakeholders, and decision-makers to understand problems and find solutions. Most of the data collection, analyses, and mapping in the 1980s were done manually or semi-manually. The author tried to find better ways to automate the tedious and error-prone mapping and data analyzing tasks he was managing. After extensive research, he realized that GIS technology was the best choice to improve the efficiency of geological and environmental mapping and data collecting, processing, editing, storing, managing, sharing, analyzing, modeling, and visualizing processes. He started to learn and use GIS in the late 1980s, with very limited computing and data storage resources.

In March 1994, the author and another geologist organized the IGCP Project 294 International Symposium in Xi'an, China, and edited a book entitled *Very Low-Grade Metamorphism: Mechanisms and Geologic Applications*, published by the Seismological Press. During this conference, the author connected with some environmental scientists and geologists from Japan and the United States, who used GIS in their studies and projects.

In early 1995, sponsored by the China Ministry of Science and Technology and Japan International Cooperation Agency, the author traveled in Japan to study some environmental sites impacted by historical mining, especially the gold mines constructed and operated after World War II. He worked there for around half a year, and had more access to GIS technology, especially the Environmental Systems Research Institute's (ESRI's) first PC version ARC/INFO software.

In fall 1996, funded by his professor's National Science Foundation (NSF) grants, the author came to the United States as a graduate student, assisting his professor in teaching and research. Besides his geology studies, he devoted a lot of time to learning GIS and computer programming. As part of this M.S. thesis, he was tasked to convert the hand-drawn geology maps of a southern California and southwestern Arizona region into digital GIS data layers and maps, and then write GIS programs to read geologic measurements from tables/databases and plot them automatically onto geology maps.

At that time, GIS technology was still at its early stages of development. People had just started to learn and use it in their disciplines. On the campus, although there were plenty of GIS experts and geology professors, there was no one who knew both GIS and geology enough to help him. As a pioneer pursuing these interdisciplinary approaches, he soon ran into many challenges and problems. Studying and experimenting hard, he overcame these challenges one by one and successfully completed the tasks. When he presented the beautiful digital GIS maps, showing colorful geologic units and structures and measurement symbols to his professor and students, they were impressed with his innovative interdisciplinary approaches and the powerful GIS technology.

In April 1998, the author presented a paper entitled "Applications of GIS in Structural Geology" to the 110th Annual Conference of the Iowa Academy of Science. In late 1998, he published the details of his research and experiments in GIS and geology, along with many GIS, Visual Basic, C++, OpenGL, and virtual reality programs in his master's degree thesis of "Computer Mapping, Data Analysis, 3D Modeling, and Visualization of the Geology in the Picacho-Trigo Mountains Area of Southern California and Southwestern Arizona." The author also published his GIS programs to an ESRI website to share with other GIS and geology users. They attracted a large number of users with many downloads, questions, and complimentary comments. His pioneering work helped his department's students and some faculties to learn and use GIS in their geology and environmental studies and research. Alongside the geology projects, the author was also selected to work as a research assistant on a natural resource management GIS project of the U.S. National Natural Resources Inventory (NRI). His knowledge and experience in GIS was further broadened through that project.

Thanks to his papers, GIS programs, and thesis, the author received multiple job offers before he graduated. In late 1998, he accepted a challenging offer from a small environmental and engineering company in New York to start up and manage their GIS department. Under his management, GIS technology was extensively used in their environmental, geology, planning, natural/cultural resources conservation, and engineering projects.

In March 2002, the author was hired to manage a large engineering company's GIS department, which for years had been plagued with issues of data, mapping, software, and hardware. With his in-depth GIS knowledge and management experience, he successfully identified and solved the problems, and quickly enabled the GIS department to operate efficiently. Since then, his GIS department has supported hundreds of projects in environmental, geology, transportation, utilities, natural and cultural resources, planning, land management, construction, national security, disaster assessment and responses, and international humanitarian aid, with more than 1000 project sites all over the United States, its territories (e.g., Guam, Northern Mariana Islands, American Samoa, Saipan, Puerto Rico, and Virgin Islands), and other countries, such as Iraq, Libya, Kuwait, Saudi Arabia, Rwanda, and Haiti.

Through these numerous and diverse worldwide projects that the author worked on and managed in his 30-year career, he has accumulated a large amount of data, knowledge, experience, and lessons. Based on them, he has written a variety of papers, and eventually this book. This book was therefore written from the unique perspectives of a GIS learner/student, a user, a senior manager and a geologist, as well as an environmental scientist.

1

General Overview of GIS

What Is GIS?

A Geographic Information System (GIS) is a computer system that helps users to collect, process, edit, store, manage, share, analyze, model, and visualize large volumes of datasets to understand spatial relationships, patterns, and trends, and make educated and sound decisions. The two key words in its name, "geographic" and "information," make GIS a unique technology, different in many ways from other traditional cartographic mapping and data management systems and applications. "Geographic" (also known as geospatial or spatial) means that GIS mainly deals with spatial data features, which are in one way or another references to locations on the earth, such as sampling location points, utility lines, and land parcel polygons. By manipulating spatial data layers (also known as themes), GIS is capable of producing maps similar to other mapping systems, such as a computer-aided design and drafting (CADD) system.

In addition to mapping capabilities, GIS incorporates tabular data (i.e., "information") and links information to spatial data features (e.g., attaching analytical results to sampling locations) to enable data analysis, modeling, and visualization for better understanding of data and making decisions. For example, by analyzing the population/demographic information, traffic, existing business, and client data layers of an area with a GIS, a business owner or manager will be able to identify the best locations for businesses, such as shopping centers, restaurants, hotels, offices, warehouses, service facilities, and so on. For an environmentally impacted area, by analyzing historical and current aerial photos, sampling results, water wells, hydrology, geology, biology, census, parcels, and other related datasets, scientists and decision-makers can better understand the site history and its current conditions; evaluate its risks to ecosystems, the environment, and human health; model or simulate its possible future impacts; and design the most appropriate remedial approaches to mitigate the situation and clean up the site.

During a GIS conference, one presentation discussed how GIS technology was used in a flooding disaster assessment and its relief efforts. After massive flooding in a south Asian region, airports and road connections to the outside world were severely damaged. It was impossible to go there quickly to survey the affected areas and accurately prepare disaster relief plans. Luckily, using social media, some local residents and officials managed to send out valuable information about the flooding, such as pictures and locations of the places where people were stranded, usable road segments, still-functioning facilities, and so on. GIS technology was used to gather and process these datasets to help United Nations and local officials to quickly assess the situation and coordinate relief efforts from outside and inside the affected region.

GIS is widely used in natural/cultural resource management and conservation, environmental investigation and remediation, geosciences, engineering, logistics, planning, transport, utilities, telecommunication, aviation/navigation, surveying, demographic studies, epidemiology, emergency management, disaster relief, humanitarian aid, forestry, agriculture, oil/gas/mineral exploration, mining, real estate, retail/warehousing/distribution, insurance, health/medical resource management, education, training, tourism, security, law enforcement, reconnaissance, military operations, and many other fields.

GIS Data

The most important and essential component of any GIS is data, which includes both spatial datasets and their associated metadata. Spatial data (also commonly referred to as geospatial data, geographic data, or GIS data) is data or information about the locations and shapes of physical features and the relationships between them. The locations and shapes of features are usually stored as coordinates, commonly known as X, Y, and Z values in a projected coordinate system, or latitude, longitude, and altitude in a geographic coordinate system. The relationships between features are stored as topology, which represents geometric properties, spatial relations, and the arrangement of the features. With stored information about the locations, shapes, and their relationships of features, spatial datasets (also known as data layers or themes) can easily be manipulated and analyzed in GIS to find useful information, patterns, and trends to make educated and sound decisions. This is the unique power of GIS.

Due to its spatial nature, a special database format, a *geodatabase*, is needed to store GIS data. Major relational database management systems (RDBMS) have tools and/or utilities to design, develop, deploy, populate, and manage geodatabases, such as IBM DB2, IBM Informix Dynamic Server, Microsoft Access, Microsoft SQL Server, Microsoft SQL Server Express, Microsoft SQL Azure, PostgreSQL, Oracle, and so on.

Metadata is the data or information that describes the spatial dataset. The prefix *meta* in Greek means "after," "along with," "beyond," "among," and "behind." It adds to the name of something references, comments, and/or other information. In order for GIS data users to precisely understand, interpret, exchange, and reuse spatial data, data production processes are documented and saved (along with other relevant information about the data) as metadata. A metadata file contains detailed information about the spatial data, such as its name, purpose, spatial domain/bounding coordinates, scale, projection/coordinate system, data originators, dates/times, keywords, use constraints, data quality/accuracy, sources, production processes, data formats, attributes/fields, legal disclaimers, distribution/access, and so on. Figure 1.1 shows an example of a metadata document. It is the first page (i.e., the data "Identification

Metadata:

- Identification_Information
- Data_Quality_Information
- Spatial_Data_Organization_Information
- Entity_and_Attribute_Information
- Distribution_Information
- Metadata_Reference_Information

Identification_Information:
  Citation:
    Citation_Information:
      Originator: U.S. Geological Survey
      Publication_Date: Unknown
      Title: DRG-24k Template
      Geospatial_Data_Presentation_Form: map
      Series_Information:
        Series_Name: Digital Raster Graphics
        Issue_Identification: 0.1
      Publication_Information:
        Publication_Place: Sioux Falls, SD
        Publisher: U.S. Geological Survey
      Other_Citation_Details:
        Earth Science Information Centers (ESIC) offer nationwide information and sales service for USGS map products and earth science publications. For additional information, contact any USGS Earth Science Information Center (ESIC), or call 1-888-ASK-USGS.
      Online_Linkage: <http://seamless.usgs.gov>
  Description:
    Abstract:
      A digital raster graphic (DRG) is a scanned image of a U.S. Geological Survey (USGS) topographic map. The scanned image includes all map collar information. The image inside the map neatline is georeferenced to the surface of Earth. The DRG can be used to collect, review, and revise other digital data especially digital line graphs (DLG). When the DRG is combined with other digital products, such as digital orthophoto quadrangles (DOQ) or digital elevation models (DEM), the resulting image provides additional visual information for the extraction and revision of base cartographic information. The USGS is producing DRGs of the 1:24,000-, 1:25,000-, 1:63,360-(Alaska), 1:100,000-, and 1:250,000-scale topographic map series.
    Purpose:
      The DRG is used for validating digital line graphs (DLGs) and for DLG data collection and revision. The DRG can help assess the completeness of digital data from other mapping agencies. It can also be used to produce "hybrid" products. These include combined DRGs and DOQs for revising and collecting digital data, DRGs and DEMs for creating shaded-relief DRGs and combinations of DRG, DOQ, and DLG data. Although a standard DRG is an effective mapping tool, its full potential for digital production is realized in combination with other digital data.
    Supplemental_Information:
      Digital raster graphic (DRG) data on CD-ROM are being produced by the U.S. Geological Survey (USGS) through an Innovative Partnership agreement with The Land Information Technology Company Ltd. of Aurora, CO. This series includes DRG's of USGS standard series quadrangle maps of the United States, its Trusts, and Territories.

AREA	PERIMETER	DRG	XMIN	YMIN	XMAX	YMAX	UNITS	SCALE	ZONE	QUAD
	51656.109	O30081G4	452137.21875	3401768.5	464157.15625	3415677.5	METERS	24000	17	Cumberland Island South

  Time_Period_of_Content:
    Time_Period_Information:
      Single_Date/Time:
        Calendar_Date: unknown
    Currentness_Reference: ground condition
  Status:
    Progress: Complete
    Maintenance_and_Update_Frequency: As needed
  Spatial_Domain:
    Bounding_Coordinates:
      West_Bounding_Coordinate:-81.49174159
      East_Bounding_Coordinate:-81.38190879
      *North_Bounding_Coordinate:*30.78776898
      *South_Bounding_Coordinate:*30.72748865
  Keywords:
    Theme:
      Theme_Keyword_Thesaurus: None
      Theme_Keyword: aerial photograph

FIGURE 1.1
An example metadata of a United States Geological Survey (USGS) digital topographic 7.5 minute quadrangle image data, also known as a digital raster graphic (DGR), which is often used as background image in GIS maps.

Information" section) of the metadata file of a U.S. Geological Survey (USGS) digital topographic 7.5 minute quadrangle image spatial data. As listed on the top of this first page, the whole metadata file contains six sections, that is, the Identification Information, Data Quality Information, Spatial Data Organization Information, Entity and Attribute Information, Distribution Information, and Metadata Reference Information. The Identification Information section contains general information about the spatial data, including data producer/originator (e.g., USGS Earth Science Information Center), publication place and date, abstract, purpose, quad name (e.g., Cumberland Island South), DRG number (e.g., O30081G4), area/perimeter of the data coverage, data units (e.g., meters), minimum and maximum coordinates, data scale (e.g., 1:24,000), UTM zone (e.g., 17), time period of the data, data production status (e.g., complete), data maintenance/update information, keywords, spatial domain/bounding coordinates, and other relevant information.

All federal agencies of the United States are mandated by Executive Order 12906 to use metadata standards endorsed by the Federal Geographic Data Committee (FGDC). The current FGDC-endorsed and widely used standard is the Content Standard for Digital Geospatial Metadata (CSDGM). Its biological extension is the Biological Data Profile (BDP). The FGDC has also endorsed the adoption of the International Organization for Standardization (ISO) series of standards (19115, 19115-2, and 19139). A transition to the ISO series of standards is already underway. Both CSDGM and ISO require metadata to be formatted in Extensible Markup Language (.xml), although many old-version metadata files are in HyperText Markup Language (HTML), TXT, Microsoft Word, and other formats.

There are a variety of free or commercial software tools with a wide range of features and capabilities to create, edit, and validate metadata files. A few well-known and commonly used metadata software tools are briefly discussed below.

Freeware/Shareware Metadata Tools

- *EPA Metadata Editor (EME)*: This is an extension to ESRI's ArcGIS ArcCatalog software product. ArcGIS users can utilize this EPA Metadata Editor to create and edit records that meet the EPA Geospatial Metadata Technical Specification and the FGDC CSDGM requirements.

- *NOAA MERMAid (Metadata Enterprise Resource Management Aid)*: Developed and maintained by the National Oceanic and Atmospheric Administration (NOAA)'s National Centers for Environmental Information (NCEI)'s Center for Coasts, Oceans, and Geophysics (CCOG), MERMAid is an online metadata editor for creating, editing, validating, storing, and exporting metadata records. It supports the FGDC CSDGM with the BDP, the Shoreline Profile, the Remote Sensing

(RS) Extension, Ecological Metadata Language (EML), and Machine-Readable Cataloging (MARC). It can be used for metadata conversion among these standards too. The NOAA recently announced that MERMAid would be discontinued as of September 30, 2016, in line with the overall trend of transitioning from FGDC CSDGM metadata standard to the ISO series of standards of 19115, 19115-2, and 19139.

- *USDA Metavist*: Produced and maintained by the USDA Forest Service, Metavist is a stand-alone desktop metadata editor for creating metadata compliant with the FGDC CSDGM metadata standard and the National Biological Information Infrastructure (NBII) 1999 BDP for the FGDC standard. The output metadata is in XML format, required by the FGDC CSDGM metadata standard.

- *USGS Online Metadata Editor (OME)*: OME is an online metadata editor for creating FGDC CSDGM-compliant metadata records by answering questions about the spatial data. It can be used to create new metadata records or upload/edit existing ones. Metadata records can be saved either online or locally to a desktop computer. This special online metadata tool is currently available to USGS staff only.

- *USGS Metadata Wizard*: This is a metadata tool that works with ESRI's ArcGIS Desktop for generating FGDC CSDGM-compliant metadata for geospatial datasets. This metadata tool is a Python toolbox for ESRI ArcGIS Desktop software, which processes geospatial datasets and creates/updates metadata records in ESRI's ArcGIS software.

- *USGS Metadata Parser (MP)*: Developed and maintained by the USGS, MP is a metadata quality control and output configuration tool. It supports Linux, UNIX, and all versions of Microsoft Windows operating systems. It works as a compiler to parse metadata, that is, checking the syntax of the metadata against the FGDC CSDGM standard and generating error reports that can be viewed with - either a Web browser or text editor. It is multilingual, supporting French and Spanish. It can be configured for the BDP and other extensions.

- *Microsoft XML Notepad*: XML Notepad is an open-source lightweight metadata tool published by Microsoft. It can be used to view, edit, and validate existing metadata records and XML documents. However, it does not create FGDC CSDGM metadata records. With a special schema package, XML Notepad can be used to validate metadata records.

Commercial Metadata Tools

- *ESRI's ArcCatalog*: With ArcCatalog, the existing metadata of a spatial data layer can be viewed, edited, and validated. For a data layer without any metadata associated with it, ArcCatalog can automatically create a basic metadata file for it with items and formats compliant with either the FGDC CSDGM metadata standard or the ISO series of standards of 19115, 19115-2, and 19139. From the information of the spatial data, ArcCatalog automatically fills in some of the items in the metadata file, such as the title, thumbnail, spatial domain/ bounding coordinates, scale, projection/coordinate system, attributes/fields, and so on. An existing or newly created metadata file can be edited easily within ArcCatalog, similar to a text editor.

- *GeoMedia Catalog*: Intergraph's GeoMedia Catalog is capable of creating, managing, and publishing metadata records compliant with FGDC CSDGM, ISO 19139, and other metadata standards. Since GeoMedia Catalog associates spatial data and metadata catalog entries together, it can automatically extract from the spatial data the information for some metadata elements, such as bounding coordinates, point and vector object counts, native data environment, entity/attributes, and so on. GeoMedia Catalog allows multiuser access to the GeoMedia Catalog metadata repository through multiple GeoMedia Catalog clients using Microsoft Access, SQL Server, or Oracle databases.

- *MapInfo Manager*: Pitney Bowes MapInfo Manager is a server-based GIS application. With MapInfo Manager, users can create and edit metadata compliant with the ISO 19115 metadata standard. Some metadata-generating processes can be automated using templates and/or performed in a bulk collection effort, referred to as *harvesting*.

There are two primary types or formats of spatial data, that is, vector data and raster data. The vector data format is more commonly used. In a GIS, vectors are geographically enabled geometrical shapes. Their primitive types include zero-dimensional points, one-dimensional lines, and two-dimensional polygons. Examples of point features are sampling locations, monitoring stations, wells, fire hydrants, traffic signs, and so on. On a small-scale map that shows a large geographic area, such as the whole world, a continent, or a large country, cities are represented as points. However, on a large-scale map that shows a small area, such as a state/province or a county, cities are displayed as polygons. Examples of line features include utility lines (power lines, water/sewer networks, oil and gas pipelines), streams, rivers, shorelines, survey paths, roads, rail tracks, and so on. However, on large-scale maps, some line features can become polygons, for example, roadways, rail tracks, easements, large rivers, and so on. Polygon features are the most common geometrical shapes on the earth's surface. Figure 1.2 is an example of a GIS map showing groundwater monitoring wells (red stars) as points, streams (cyan) as lines, and a volatile organic compound (VOC)-contaminated groundwater area (shaded in yellow) as a polygon. The changes of the chemical concentrations in the groundwater during the 9-year

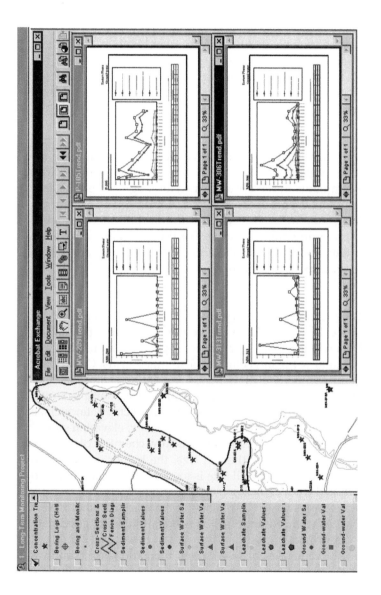

FIGURE 1.2

A GIS map showing monitoring wells (red stars) in a volatile organic compounds (VOCs)-contaminated area (shaded in yellow) inside the formerly used Naval Air Station in Brunswick (NASB), Maine. The NASB was closed in 2011. For each well, the chemical analytical results of the groundwater samples collected over a 9-year sampling period was displayed as a line graphic and a data table. On the top left of the graphic is the well ID (e.g., MW-209, MW-313, MW-306). The color lines represent the chemicals. The values on the vertical axis of the graphic are concentrations, while the numbers on the horizontal axis of the graphic are sampling events (around 25 events) over the 9-year sampling period, with around 2–3 sampling events each year. The chemical concentration line graphics and data tables were generated with Microsoft Access database or Excel, and converted into PDF files, which were linked to the monitoring wells on the GIS map.

sampling period were displayed as a line graphic and data table for each well. The line graphics and tables were created with the Microsoft Access database or Excel and converted into PDF files. The PDF files were linked to the monitoring wells on the GIS map. When a monitoring well is clicked, detailed information about the well shows up on the GIS map, including the well ID, its coordinates, total depth, depth to groundwater, permit info, construction date, construction diagram, sampling information, chemicals analyzed, chemical concentration data tables/line graphics, and much more other relevant information. The chemical concentration values of the wells at each sampling event and their changing trends over a certain sampling period were used in designing and optimizing the water treatment system of the contaminated groundwater body, also called a *plume*. The line graphics show clearly that overall the chemical concentrations decreased during the sampling events although there were wild variations at some sampling events. This changing trend indicated that the water treatment system worked well in reducing the contaminants in the groundwater, although some problems occurred at a few sampling events. As soon as the problems were discovered by analyzing the data, they were corrected by tuning the water treatment system and taking other mitigating measures. For this environmental remediation project, GIS data analysis and modeling played a significant role in helping scientists, engineers, and decision-makers to understand problems and find the best solutions. It will be discussed in more detail in the following chapters.

Raster data is a type of digital imagery that is composed of an array of same-sized cells, also called *pixels*. Each cell contains a value representing the information of the whole cell area or its center-point location. The cells are organized into regular rows and columns, also called grids. Commonly used raster data examples include digital pictures, aerial photos, satellite images, remote sensing images/data, geophysical survey data, scanned documents, and statistically analyzed data from certain vector data, such as population density, anomaly density, fracture line density, elevation slope/aspect/shaded relief, surfaces of elevation, temperature, chemical concentration, and so on. Usually, the smaller the cell size, the higher quality (also called resolution) the raster data. If a cell size is too large, the imagery is too coarse to display detailed information or small features. However, higher-resolution raster data requires larger disk space for storage, and takes longer to access and process. A balanced approach is needed to ensure good data quality and system performance.

There are three major categories of raster data: thematic raster, continuous raster, and digital pictures (also including scanned images). A *thematic raster* (also known as *discrete, categorical,* or *discontinuous* raster), represents a single topic or theme, such as geologic units, soil types, land uses, vegetation types, and so on. Figure 1.3 is an example of a land use (also known as land cover) thematic raster map of a project site in Florida. In a thematic raster, a value assigned to a cell represents the information of the whole cell area,

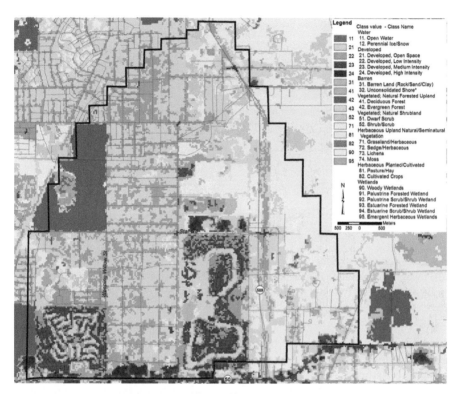

FIGURE 1.3

A land uses GIS map of a project site in Florida. This land uses image is a thematic raster. Each land use type has its unique cell value and its clear boundary. The land use types displayed in this raster image can be analyzed and converted into vector land use vector polygons.

and features have their clear boundaries, such as geologic units, land uses, vegetation types, and so on. There is no transition between features.

In a continuous raster dataset, cell values vary continuously to form a surface, such as elevation surface, population density, chemical concentration, distance from an object, aerial photos, satellite images, and so on. The value assigned to a cell is represented at its center point, instead of the whole cell. Figure 1.4 is a three-dimensional topographic elevation surface map, generated from a U.S. Geological Survey (USGS) 7.5 minute quadrangle Digital Elevation Model (DEM) dataset, which has now been replaced by the newer version of the National Elevation Dataset (NED). Overlaid on top of this 3D topographic elevation surface map are the layout of the formerly used Naval Air Station Brunswick (the black vector lines) and a VOC-impacted area (the pink polygon in this figure and yellow polygon in Figure 1.2), where the groundwater underneath it was contaminated by VOCs. This GIS map is useful in understanding the general topographic setting of the project site and its surrounding areas. As discussed earlier, a groundwater treatment system was designed and installed to remove or reduce the VOC concentrations

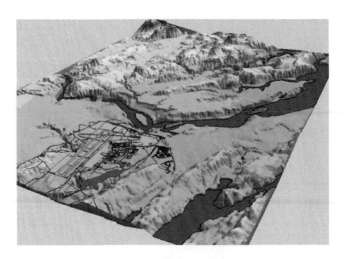

FIGURE 1.4
A 3D topographic elevation map generated from an USGS 7.5 quadrangle DEM dataset, showing the layout of the formerly used Naval Air Station in Brunswick (NASB) and a volatile organic compound (VOC) impacted area (pink polygon). The NASB was closed in 2011.

from the groundwater plume below a permitted level, compliant with the relevant federal and state laws and regulations, so that it would not impact the river passing its north boundary, and the ocean in the south. 3D models of this contaminated groundwater plume and an animation were generated from the dataset stored in a GIS database. They will be discussed in more detail in the 3D modeling and visualization chapter (Chapter 5).

Coordinate Systems and Projections

A coordinate system is a reference system that determines the position of a point or other geographic element on the earth's surface. There are two commonly used coordinate systems in the GIS world, that is, geographic coordinate system (GEOGCS or GCS), also known as the "spherical" or "global coordinate system," and the projected coordinate system (PROJCS), which projects or flattens the earth's spherical surface onto a two-dimensional Cartesian coordinate plane or a map sheet, using various mathematical models and mechanisms.

A geographic coordinate system uses a three-dimensional spherical surface to define locations on the earth, which includes angular units of measures (i.e., latitude and longitude in decimal degrees), a prime meridian, and a datum. Latitude and longitude are angles measured from the earth's center to a location on the earth's surface, referencing the equator line and the defined prime meridian. The latitude of the equator line is 0° and Latitude of the North or South Pole is 90°. Latitudes north of the equator are positive numbers (0°–90°) or designated with the letter "N", for example, 90°N for the North Pole, while latitudes south of the equator are negative numbers (0° to −90°) or designated with the letter "S", for example, −90°or 90°S for the South Pole. The longitude of the prime meridian is 0°. The longitudes east of the prime meridian are positive numbers (0°–180°) or designated with the letter "E" (0°–180°E), while longitudes west of the prime meridian are negative numbers (0° to −180°) or designated with the letter "W" (0°–180°W). The *international prime meridian* is the meridian of the British Royal Observatory in Greenwich, east of London, England. The *antipodal meridian* of Greenwich is 180°W or 180°E. The imaginary latitude and longitude lines divide the world into an angular grid system as shown in Figure 1.5. A pair of latitude and longitude coordinates locates the position on the surface of the Earth. The geographic coordinate system is widely used in GPS, mobile GIS systems, navigation and aviation maps, small-scale world, regional, and country maps, and others.

In a projected coordinate system, a geographic feature on the Earth's spherical surface is projected onto a two-dimensional flat plane, transforming from its original spherical coordinate system (latitude and longitude) into a planar coordinate system (easting and northing), using a suitable mathematical model. Since it is on a flat surface, a projected coordinate system has lengths, angles, and areas. Locations are identified by easting (X) and northing (Y) coordinates, which are distance measurements from the defined origin of a grid system. Since ancient times, people have been exploring and experimenting countless mathematical theories and methods to perfectly project the Earth's spherical surface onto a flat map sheet without any distortions in the shape, area, distance, or direction of the data. Unfortunately, it is proven to be impossible to achieve that goal. In every projection, at least some part of the mapped region is distorted in its shape, area, distance, or direction.

A good example is the Christopher Columbus voyages between Spain and the Americas from 1492 to 1503. He originally planned to find a short western sea route across the Atlantic Ocean between Spain and Asia for trade purposes. However, referring to the primitive Ptolemy's Geographia theories of the earth's surface and the distorted world map, Christopher Columbus significantly underestimated the distance from Europe to Asia across the Atlantic Ocean, and landed in the American continent by mistake. He thought that he had reached his destination of India, as originally planned and sponsored by the Spanish Crown of Castile. Christopher Columbus even kept on denying his mistake, and therefore his credit for such a great accidental discovery of the New World for a few years afterwards.

Although all projections have certain types of distortions, some projections can minimize the distortion of one or two properties, such as distance, area, shape, or direction. With newer and better technologies, understanding

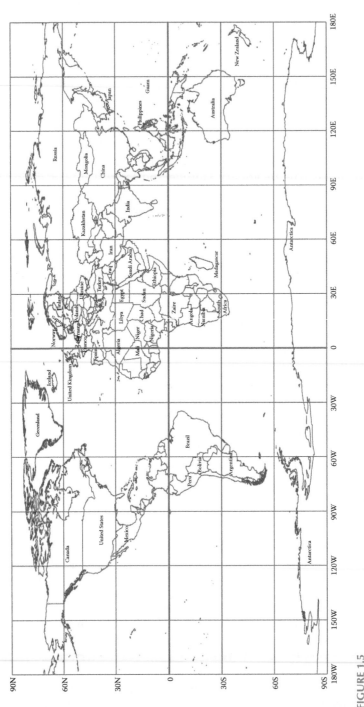

FIGURE 1.5
A simplified world map overlaid with latitude and longitude lines. The numbers are in decimal degrees and the interval is 30°. The 0° longitude line is the prime meridian, and the 0° latitude line is the equator.

of the earth has significantly improved. Traditional projections have been improved, and new projections have been developed for various purposes. Nowadays, there are hundreds of projections available to choose from. It can be a challenging task to identify the most appropriate projection for a dataset, depending on multiple factors.

Based on the types of geometric shapes that are used to transfer features from a sphere or spheroid to a plane, projections can be grouped into three major families, that is, conic, cylindrical, and planar. Based on their properties, projections can also be classified into three major types, that is, conformal, equal area, and equidistant. Conformal projections preserve angles locally, with local shape less distorted, but area distorted significantly, for example, the Mercator, Lambert Conformal Conic, and Stereographic projections. Equal area projections, as the name implies, preserve the areas of polygon features, but with other properties, such as shape, angle, and scale distorted, for example, the Lambert cylindrical, Albers conic, and Flat-Polar Quartic projections. Equidistant projections preserve the distances between points better, but scale may not be maintained accurately, for example, the equidistant conic and equirectangular projection. There is another projection type called "compromise," which is something in between the major projection types, and designed to achieve a good balance to reduce overall distortion, for example, the Web Mercator projection. Each map projection has a set of parameters that specify the details of the projection, such as datum, origin, unit, scale factor, prime meridian, central meridian, false easting, false northing, and so on.

Since the Earth's surface, with mountains and valleys, is not a perfectly smooth sphere, a mathematical model, known as a datum, is needed to fit the earth to an ellipsoid. It specifies the origin and orientation of latitude and longitude lines, among other things. Since there are so many estimates of the size and shape of the Earth, many datums have been developed. Globally, commonly used datums are the WGS1984 (World Geodetic Survey, 1984), Clarke 1866 ellipsoid, GRS80 (Global Reference System, 1980), and the ITRF97 (International Terrestrial Reference Frame 1997). Some global datums are widely adopted and used regionally, and even locally. For example, the China Geodetic Coordinate System 2000 (CGCS2000) is based on the ITRF97 datum. Regionally, on the North America continent, two local datums were developed, that is, the NAD1983 datum (North American Datum, 1983), which is similar to WGS84, and the NAD1927 datum (North American Datum, 1927), which is similar to Clarke 1866. In Europe, the ED50 (European Datum 1950) was developed. In Africa, the Arc 1950, Arc 1960, and Adindan datums were developed and are used by different nations. There are a variety of other local datums, such as the Potsdam of Germany, China National Geodetic Datum 2000, GDA 1994 of Australia, Tokyo of Japan, Afgooye of Somalia, South Asia datum of Singapore, Campo Inchauspe of Argentina, Cape datum of South Africa, CH 1903 of Switzerland, Chatham 1971 of New Zealand, Corrego Alegre of Brazil, Djakarta of Indonesia, Nahrwn of Saudi Arabia, Old Egyptian datum of Egypt, and so on.

In North America, commonly used projections include Albers, Lambert, and Mercator. In the 1940s, the United States Army Corps of Engineers (USACE) developed the famous and widely used UTM (Universal Transverse Mercator) coordinate system, which divides the Earth surface between 80°S and 84°N latitudes into 60 zones, as shown in Figure 1.6.

Depending on the location in the world, different datums can be used in the UTM coordinate system, such as Clarke 1866 ellipsoid, NAD83, WGS84, and so on. The UTM coordinate system is a variant of the conformal Mercator projection, which preserves local angles and shapes but distorts distance and area. This coordinate system is more suitable for a large-scale map with mapping areas/regions mainly inside one UTM zone or two. If the region is too large, covering multiple UTM zones, it will be complicated to re-project maps and datasets between different zones, and its area could be distorted significantly.

In the United States, each state or territory (such as Puerto Rico and US Virgin Islands), has its own State Plane Coordinate System (SPCS), with one or multiple zones, oriented roughly horizontally (east–west) or vertically (north–south) following the general overall shape of the state, as shown in Figure 1.7. Most SPCS zones are based on either a transverse Mercator projection or a Lambert conformal conic projection, depending on the shape of the state or territory. A Lambert conformal conic projection is usually used for states stretched in an east–west direction because a conic projection preserves accuracy along the east–west axis better. In contrast, a transverse Mercator projection is typically used for states elongated along a south–north direction because a cylindrical projection preserves accuracy along the south–north axis better. Oblique Mercator projection is used for eastern Alaska due to its unique diagonal shape. North Carolina's Department of Transportation (NCDOT) initiated the efforts in developing its state plane systems in 1933. It soon became popular and was adopted by other states and territories. Originally, an SPCS was based on NAD1927 datum. Later on, it was switched to the newly developed and more accurate NAD1983 datum.

Every projection was designed for a special purpose and is good and useful in one way or another. However, there is no such a thing as a perfect projection available today and there will not be one in the future that can be universally used for all maps and datasets. Therefore, it is important to understand the detailed characteristics of projections and to select the most appropriate or suitable projection for the interested area to reduce the distortion(s) as much as possible.

Every GIS dataset must have a defined coordinate system for it to be displayed correctly in a map, and integrated with other data layers to perform various data analysis operations. For most GIS datasets, the parameters and other information about their coordinate systems are stored in their associated projection files (".prj") for vectors, and world files for rasters, for example, ".jpw" files for JPEG images, ".tfw" files for TIFF (Tagged Image File Format) images, ".sdw" files for multiresolution seamless image database

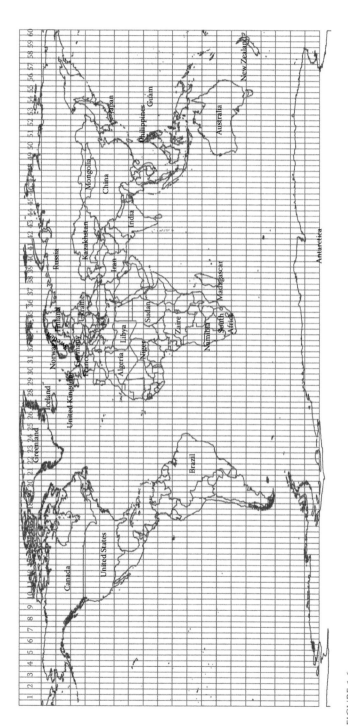

FIGURE 1.6

A map showing the 60 UTM zones (blue lines with red labels) covering the whole world, developed by the United States Army Corps of Engineers (USACE) in the 1940s.

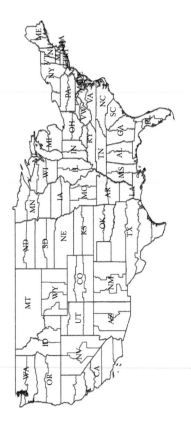

FIGURE 1.7
A simplified map of the United States, with NAD1983 state plane coordinate system (SPC) zones (red lines).

```
PROJCS["WGS_1984_UTM_Zone_55N",GEOGCS["GCS_WGS_1984",DATUM
["D_WGS_1984",SPHEROID["WGS_
1984",6378137.0,298.257223563]],PRIMEM["Greenwich",0.0],UNIT
["Degree",0.0174532925199433]],PROJECTION
["Transverse_Mercator"],PARAMETER
["False_Easting",500000.0],PARAMETER
["False_Northing",0.0],PARAMETER
["Central_Meridian",147.0],PARAMETER
["Scale_Factor",0.9996],PARAMETER["Latitude_Of_Origin",0.0],UNIT
["Meter",1.0]]
```

FIGURE 1.8
An example projection (".prj") file of a vector GIS dataset or feature class.

(MrSID) images, and so on. Some formats of raster data, such as GeoTIFF, IMAGINE, grids, bip, bil, and bsq, store their georeferencing information in the header of the image files, so world files are not needed for these special raster datasets. However, no vector dataset has this option. Therefore, a vector data must have projection information associated with it.

Figure 1.8 is an example projection (".prj") file, which tells a GIS system that the spatial data associated with it is in the projected coordinate system of UTM zone 55 North, WGS 1984 datum, and based on the GCS of WGS 1984, with datum of WGS 1984. It also contains other detailed information about the projection (transverse Mercator), spheroid, linear parameters (false easting and false northing), angular parameters (prime meridian, central meridian, latitude of origin), scale factor, units, and so on.

In a raster dataset, cells are organized into regular rows and columns. Each cell has a row and column number. Its associated world file stores the information used to transform and convert the image coordinates into real-world coordinates. Compared with vector projection files, raster world files are much simpler and their contents are in a similar format. Figure 1.9 is an example of an ".sdw"-type world file of a USGS 7.5 minute topographic quadrangle (also known as DGR) image in MrSID compressed format. The first line is "X" scale, that is, the dimension of a pixel in map units in "X" or easting direction. The second and third lines are rotation terms. The fourth line is negative of "Y" scale, that is, the dimension of a pixel in map units in "Y" or northing direction. The last two lines are translation terms, that is, X and Y map coordinates of the center of the upper-left pixel.

```
            2.438400000000000
            0.000000000000000
            0.000000000000000
           -2.438400000000000
       426447.760084074398037
      3194396.661732787732035
```

FIGURE 1.9
An example of an ".sdw"-type world file of an USGS DGR image in MrSID compressed format.

Spatial Data and Metadata Standards and Relevant Guidelines

In order to standardize GIS data developments and facilitate data sharing, various standards and guidelines for spatial data and metadata have been developed. In the United States, the most widely used enterprise-level standards are the Defense Installation Spatial Data Infrastructure (DISDI) Group's Spatial Data Standards for Facilities, Infrastructure, and Environment (SDSFIE), and the Federal Geographic Data Committee's (FGDC) Content Standard for Digital Geospatial Metadata (CSDGM), Spatial Data Transfer Standard (SDTS), Geospatial Positioning Accuracy Standards, Geographic Information Framework Data Standard, Wetlands Mapping Standard, United States Thoroughfare, Landmark, Postal Address Data Standard, Federal Trails Data Standard, Coastal and Marine Ecological Classification Standard, and so on. In 1992, the Tri-Service CADD/GIS Center led by the U.S. Army Corps of Engineers originally developed the SDSFIE. In 2006, SDSFIE was completely re-engineered by the DISDI Group and its current version (3.1) is now a family of seven standards. The FGDC was established in 1990 and sponsored by the U.S. Geological Survey.

There are many other national and international spatial data and metadata standards, such as the Canadian Geospatial Data Infrastructure (CGDI) standards, ISO 19115 Geographic Information Metadata, ISO 19139 Geographic Information Metadata XML schema implementation, and various other ISO/TC211 standards.

Besides international and national standards, some industries and agencies have their own spatial and metadata guidelines. For example, the U.S. Army Corps of Engineers developed a Data Item Description (DID) WERS-007.01, "Geospatial Information and Electronic Submittals, 4/8/2010," and several engineering manuals, such as the EM 1110-1-2909, "Geospatial Data and Systems, 9/1/2012" and the EM1110-2-6056, "Standards and Procedures for Referencing Project Evaluation Grades to Nationwide Vertical Datum's, 12/31/2010."

GIS History

GIS is a relatively new (around 50 years) technical field or discipline, branching out from traditional cartography, which has a long history of thousands of years. For various purposes, such as inventorying lands, forestry, water resources, facilities, and infrastructures; zoning for political or taxing reasons; mineral and oil/gas exploration; mining; fighting wars; finding routes for trades, development, and construction; navigating on land and seas; assessing disasters and planning responses; and so on, ancient people drew features on the earth onto maps to show their locations, shapes, and relationships with labels (also known as annotations), to help users understand the features, count them, measure their distances and areas, and make plans or decisions.

A good example is the mapping of the ancient Chinese Silk Road starting from the Western Han Dynasty's capital city of Chang'an (now Xi'an) in the east, going westward all the way to Europe, passing western China regions, India, Kazakhstan, Tajikistan, Uzbekistan, Afghanistan, Iran, Russia, Turkey, Greece, and other countries, and finally reaching the Roman Empire. The Silk Road used to be a great East to West trade and cultural exchange route passage, initiated in the second century BC (around 139–129 BC) by the adventure of a Chinese diplomat and general in the Western Han Dynasty, Zhang Qian, who served as an imperial envoy to the world outside of China. The trading route was named the Silk Road in 1877 by Ferdinand von Richthofen, a German geographer, due to the fact that Chinese silk was the most popular product transported and traded along the route. The trading and exchange activities on the Silk Road lasted a very long time. The famous Italian explorer and merchant Marco Polo traveled the Silk Road in 1271, and became a friend of the then emperor of the Yuan Dynasty, Kublai Khan, who was the grandson of the great Genghis Khan. The emperor even assigned Marco Polo to govern a Chinese region for a few years before he returned to Venice in Italy in 1295. His book (*The Book of the Marvels of the World,* also widely referred to as *The Travels of Marco Polo,* published in 1300) inspired Christopher Columbus to find a short sea route to the East across the Atlantic Ocean, because it became more and more dangerous to travel the land Silk Road due to bandits and conflicts in the late 1400s and 1500s. During its 1500-year history, a variety of individual trading routes were opened and traveled along the Silk Road passage (also called Silk Road corridor), which is why the Silk Road is also called Silk Routes. They are generalized into four routes going overall from the East (Chang'an) to the West to the Middle East, North Africa, and Europe, and southwest to India and other Asian countries. Some westward routes reached all the way to the Roman Empire, while others covered only some of the regions.

There are countless ancient sketches, drawings, and maps depicting the various trading routes; political boundaries and jurisdictions; natural, cultural, and other relevant features for navigation, management, maintenance, protection, military, and other purposes. Some of the drawings look sophisticated, while some of them appear very primitive. Figure 1.10 is a map showing the generalized approximate Silk Road passage (the red dashed lines), generated from and based on historical drawings and documents. The southern routes to India and other Asian countries were not included in this map.

Modern cartographers perform roughly the same mapping tasks as their ancient ancestors did hundreds or thousands of years ago. Ancient and modern maps look similar and function roughly the same way too. The only differences are that modern maps are much more accurate than ancient

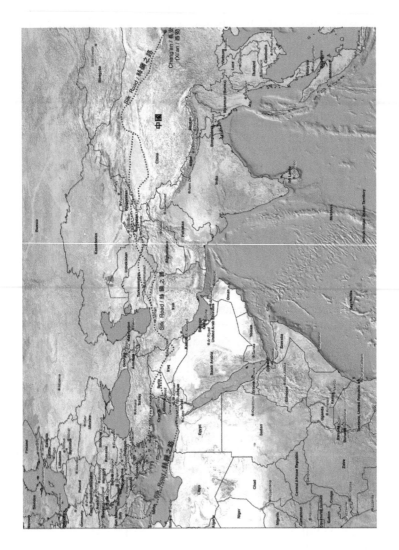

FIGURE 1.10

A map showing the generalized approximate northern Silk Road passage, also called Silk Road corridor. The Silk Road actually consists of a variety of individual trading routes going generally from the ancient China's capital of Chang'an, now Xi'an, in the East to the Roman Empire in the West, passing western China, Kazakhstan, Tajikistan, Uzbekistan, Afghanistan, Iran, Russia, Turkey, Greece, and other countries. There were also some southern routes gong to India and other Asian countries, not shown on this map. Although some routes went all the way from Chang'an to the Roman Empire, other routes covered only some of the regions.

maps, and map-production processes become more efficient due to better surveying, drafting, printing, and related technologies.

However, primitive or modern, they are all static maps, composed of geographic features of points, lines, and polygons, with limited labels (mainly names, descriptions, and/or measurements) next to them. Other than labels, users are not able to get any other information about the features from a traditional map, such as the pavement material and date of a road, its detailed designing information, current physical conditions and projected maintenance schedules, its widths and slopes at different sections, and so on. A military commander needs these types of data or information to plan their winning strategies. Transportation officials need these data to decide whether and when to maintain the road, improve it, or totally re-build it, and calculate the costs. Engineers and scientists need these data to conduct feasibility studies and design the road work. The surveyors might have collected these data about the road segments, but simply do not have the tools to store and connect these useful datasets to the features on a traditional map and communicate them to the end users. There is a disconnection between the people who have the data and the people who need them. Also, these valuable data might be lost if they are not stored and managed correctly. It is a waste of time and money if the collected datasets are lost, or not communicated to the right people to make good use to them.

Also, it is hard to visualize and analyze the relationships between data layers on a traditional paper map. For example, before designing and constructing a proposed road segment, studies on its environmental, ecological, cultural, and/or social impacts have to be conducted, reviewed, and approved by relevant regulators and stakeholders. Among other things, scientists have to study the wetland, streams, rivers, habitats, soils, geology, water wells, parcels, buildings, population/demographics, businesses, utilities, existing road network, and other things in the proposed road corridor. They need something more than a paper map to see not only these features displayed, but also know the detailed information about them and analyze their relationships. Furthermore, on small-scale maps, such as regional maps, it is impossible to show all the needed features and there is not enough space to label them all. Clearly, a new and better technology is needed, which can connect geographic features with their detailed information (also known as tabular data) so that it can not only generate high-quality and accurate maps, but also edit, analyze, model, and visualize data to support decision-making, designing, and other related tasks.

There is no doubt that many pioneers had thought about this new technology, explored it, and experimented on it. Some people failed, but some made certain progress in the right direction. Therefore, it is hard to identify the people and/or organizations who first invented this new geospatial technology—Geographic Information System (GIS).

However, in the GIS world, it is widely accepted that the establishment of the Canada Land Inventory (CLI) in 1962 was a significant milestone on the GIS road map. Originally, CLI was planning to produce more than one thousand maps inventorying national land uses and productive resources of Canada, using traditional manual cartographic methods. Due to the huge territory size of Canada, it was certainly nothing else but a daunting tedious and costly task. Roger Tomlinson, a young geographer of the Spartan Air Services of Ottawa, convinced the head of CLI, Lee Pratt, to automate the mapping process and perform some map analysis with computers, which eventually led to the birth of the Canada Geographic Information System (CGIS), a primitive GIS. It was used exclusively by the Canadian government until the 1990s. Since CGIS was not commercialized or open to the public for further enhancement, unfortunately, CGIS did not evolve into a modern mainstream GIS software product. Although there are other claims, Roger Tomlinson and the CGIS are credited with the start of modern GIS technology. They inspired many people and organizations to get involved, develop, and improve this new technology, such as Edward Horwood, Howard Fisher, Ian McHarg, Betty Benson, the Ford Foundation, IBM, the Laboratory for Computer Graphics at Harvard, the U.S. Census Bureau, the U.S. Geological Survey, Intergraph, MapInfo Corporation, ERDAS, Autodesk, the Environmental Systems Research Institute (ESRI), and many more.

Another very important milestone in GIS history is the founding of the Environmental Systems Research Institute (ESRI), a small consulting business, focusing on land use planning, by Jack and Laura Dangermond in 1969. In 1982, ESRI launched a command-line-based GIS software product, ARC/ INFO, written in FORTRAN. Although the software was useful, it was hard to remember and understand the commands and their parameters. It was also tedious and time-consuming to key in commands and their parameters manually. Also, due to computing hardware limitations in the early 1980s, such as slow processers and small disk capacities, it took a long time to edit GIS datasets, and process, analyze, and model them. A significant improvement occurred in 1986, when the ARC Macro Language (AML), a proprietary high-level algorithmic language, was released with ARC/INFO version 4.0. With that, commands could be written in AML programs and executed automatically, with fewer inputs from users in the process. It could also support very simple graphical user interfaces (GUI), such as menus and forms. The ARC Macro Language made ARC/INFO GIS software much easier to use and more productive, and thus it started to gain popularity. Figure 1.11 is an example AML program, designed for ARC/INFO to convert multiple raster images from UTM Zone X with datum NAD1927 to UTM Zone X with datum NAD1983 automatically. In a sequence, this AML program converts a raster image to a grid format file, opens a projection file (i.e., the "utm27-utm83.prj" text file), reads the projection information and instructions, projects the grid to UTM Zone X with NAD1983 datum, and finally converts the projected grid back to a raster image. When it finishes one image file, it repeats/loops this whole process for the next image, until all the files are processed. This AML program significantly improves productivity of this kind of tedious

```
/**** LOOPING AML *******************************

&S FILE = [RESPONSE 'ENTER FILE NAME']

/* ERROR CHECK
&IF [NULL %FILE%] &THEN
  &RETURN NO FILE NAME ENTERED.

&S FILEUNIT := [OPEN %FILE% STATUS -R]

&IF %STATUS% NE 0 &THEN
  &RETURN ERROR OPENING FILE.

/*READ FILE
&S RECORD := [READ %FILEUNIT% READSTAT]

/* **** PLACE ANY COMMANDS NOT TO BE REPEATED HERE ***
/* **** i.e. MAPEXTENT, PAGESIZE, VARIABLES, ETC. ***

/* START LOOP
&DO &WHILE %READSTAT% NE 102


/***** PLACE WHAT WILL BE REPEATED HERE **********************************

imagegrid %record% %record% %record%.clr
project grid %record% %record%n83 /gislib/amls/projection_files/utm27-utm83.prj

/***** LOOP TO TOP ***************************************************

  &S RECORD := [READ %FILEUNIT% READSTAT]

&END

/* CLOSE FILE UNIT
&S CLOSESTAT := [CLOSE %FILEUNIT%]

&RETURN
```

FIGURE 1.11
An AML program for ARC/INFO to project multiple GIS files from one coordinate system to another automatically.

data processing work and prevents human errors, because it only prompts a GIS operator to key in the names of the image files to be projected. No other human inputs are needed. Also, this AML program could easily be customized to project vector and raster datasets into any other projections automatically by writing the names of the files to be projected either into the program itself or in a separate text file, and modifying the projection file. A GIS operator could leave this type of program running automatically while he or she works on something else, or during nonworking time.

When computer graphic user interfaces (GUIs), such as graphics icons, menus, and pointing devices, became mainstream in the desktop computing environment in the 1990s, ESRI launched a GUI-based desktop version GIS software product, named ArcView, in 1991. ESRI also offers a variety of extensions to enhance and expand its functionalities for different and special purposes. With GUI, ArcView was much easier to learn and use than the command-line based ARC/INFO GIS software product. Also, with its more user-friendly Avenue program, ArcView quickly became a very popular GIS software product worldwide. Many GIS professionals all over the world wrote Avenue programs to customize, enhance, and expand ArcView functions for their special needs, and even posted their programs on ESRI's designated website for other users to download and use for free. ArcView helped ESRI to become the most dominant GIS software vendor since the 1990s. Since then, ESRI's GIS software products and data formats (such as shapefile and geodatabase) have been accepted as the industry standards and used by various governments, organizations, industries, and educational and research institutions worldwide.

Figure 1.12 is a small portion of an Avenue program, written for ArcView to read a large quantity of geologic strike/dip measurements from data tables or databases, and use them to plot strike and dip symbols onto geology maps automatically. It not only automates the geologic mapping process, but also prevents human errors in reading so many measurement values from data tables/databases and plotting them manually, which is a very tedious, time-consuming, and error-prone task.

Figure 1.13 is a computer screenshot, showing ArcView was executing the above Avenue program in progress. It was reading the coordinates of the locations of the strike/dip measurements and their values, plotting their locations as points (i.e., the blue dots), and finally converting them into the "T" looking strike/dip symbols (in black) and labeling them with dip measurement values in 0°–90° range. On geologic maps, strike/dip measurements are important information, which helps geologists to study geologic structures and tectonic mechanism.

In 1997, the ESRI International User Conference in San Diego, California, attracted 7000 GIS users from all over the world. In contrast, only 16 people attended ESRI's first humble user conference, in 1981. Since 2004, ESRI have released a family of powerful ArcGIS Desktop, mobile, and enterprise-level server platform software products, added a development framework, and launched the popular ArcGIS Online, a cloud-based mapping and data analysis system, which further secure ESRI's dominant status in GIS software and data market.

During the more than half a century long GIS history, due to the continuous efforts of countless individuals and organizations, and fast advances in computing and communicating technologies, useful open-source and commercial GIS software products became available in the late 1970s and early 1980s. They were significantly improved and promoted for wide use in governments and across industries in the 1990s. Entering the twentieth century, GIS industry incorporated Internet and web technologies into their applications to allow users to perform GIS work and communicate maps and data online. Nowadays, GIS technology is virtually everywhere and used by millions of people with access to the Internet. GIS technology is no longer limited to GIS professionals who have the required software, hardware,

```
                        strikdipfromtable.ave
'**********************************************************************
'Plotting strikes and dips from a table
'Created by Bai Tian, Iowa State University, July 23, 1998
'**********************************************************************
'Join or link tables
joinLinkTables = MsgBox.YesNo("Join/Link tables before plotting?", "Join/Link
tables", true)
if(joinLinkTables) then
toTableName = MsgBox.Input("Enter name of the joint to table:", "", "")
toTable = av.GetProject.FindDoc(toTableName)
if(toTable = nil) then
   exit
else
   toVTab = toTable.GetVTab
end
fromTableName = MsgBox.Input("Enter name of the joint from table:", "", "")
fromTable = av.GetProject.FindDoc(fromTableName)
if(fromTable = nil) then
   exit
else
   fromVTab = fromTable.GetVTab
end
'Define the common field
joinField = MsgBox.Input("Enter the common field:", "", "")
toField = toVTab.FindField(joinField)
fromField = fromVTab.FindField(joinField)
if((fromField = nil) or (toField = nil))then
   exit
else
'Join or link the tables
'toVTab.Link(toFIeld, fromVTab, fromField)
toVTab.Join(toFIeld, fromVTab, fromField)
end

else
  MsgBox.Warning("The table does not exits", "", "")
  exit
end

'Read x y coordinates, strike, dip data from the table
theView = av.GetActiveDoc
theDpy = theView.GetDisplay
theGList = theView.GetGraphics
pointList = theView.GetObjectTag
if(nil=pointList) then
  pointList = List.Make
  theView.SetObjectTag(pointList)
end

'Open "table1.txt" which has x y strike and dip fields
aTableName = MsgBox.Input("Enter name of the data table:", "", "")
theTable = av.Getproject.FindDoc(aTableName)
theVTab = theTable.GetVTab
xField = theVTab.FindField("x")
yField = theVTab.FindField("y")
strikeField = theVTab.FindField("Trnd")
aLabelField = theVTab.FindField("Plng")

for each rec in theVTab

   aX = theVTab.ReturnValue(xField, rec)
   aY = theVTab.ReturnValue(yField, rec)
   aPoint = Point.Make(aX, aY)
```

FIGURE 1.12

A small portion of an Avenue program for ArcView to read strike and dip measurements from large data tables or databases and plot their symbols onto digital geologic maps automatically.

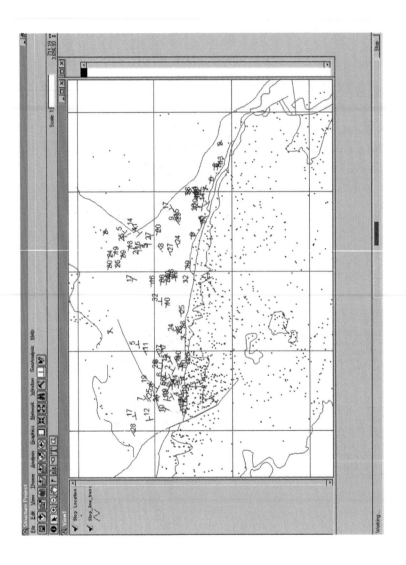

FIGURE 1.13

A computer screenshot of ArcView GIS running an Avenue program, plotting strike/dip symbols (the black "T" looking symbols) onto a geologic map, and labeling them with dip measurements. The longer lines of the symbols represent strikes, oriented based their measurements. The shorter lines of the symbols represent dips which are always perpendicular to the strike lines they associated with. The blue points are locations of the strike/dip measurements, which would eventually be converted into strike/dip symbols after their measurement values being read and processed by the program.

datasets, knowledge, and experience. Many people use GIS technology in one way or another (mainly through internet), with or without knowing it, such as the famous Google Maps, OpenStreetMap, the Global Disaster Alert and Coordination System of the UN and the European Commission, UNICEF's Progress for Children web maps, the U.S. Central Intelligence Agency's (CIA) World Fact Book, the U.S. Census Bureau's interactive population maps, the U.S. National Oceanic and Atmospheric Administration's (NOAA) interactive weather maps, and so on.

Major GIS Software Products

Major Open-Source GIS Software Products

Since 1978, numerous open-source GIS software products have been developed by various government agencies, organizations, enthusiastic individuals, and groups. As its name indicates, the source code of an Open Source Software (OSS) is available for anyone to study, modify, use, and distribute either totally free or with some licensing arrangements. The commonly used programming languages include FORTRAN, C, C++, Python, Java, ".NET," and UNIX shell. While it is virtually impossible to review all open-source GIS applications, systems, and projects, a few well-known and widely used ones are selected and briefly described below.

- *Geographic Resources Analysis Support System (GRASS, or GRASS GIS):* Under the GNU General Public License, GRASS GIS is a free open-source GIS software suite. It was originally developed for land management and environmental planning, in 1982, by the U.S. Army Construction Engineering Research Laboratory (USA-CERL), a government agency inside the U.S. Army Corps of Engineers. Since then, it has been continuously enhanced by the collaborative efforts of many government agencies, universities, and private companies worldwide. Its current version is the GRASS GIS 7, released in 2015. It contains more than 350 core modules, which support data management and analysis, image processing, graphics and map production, spatial modeling, and 3D visualization. GRASS can be either used as a stand-alone GIS application or integrated with other software packages such as Quantum GIS (QGIS) and Geostatistical R Package as their backend applications. It is one of the most well-known open-source GIS software packages, used by many academic and research institutions, governmental agencies, environmental consulting companies, and commercial and other organizations around the world. GRASS is also a founding member of the Open Source Geospatial Foundation (OSGeo), created to support the collaborative development of open-source geospatial software, and promote its widespread use. Powered by over 350 modules, the major GRASS GIS features include rendering maps and images on monitor and paper; manipulating raster and vector data including vector networks; processing multispectral image data; and creating, managing, and storing spatial data. GRASS GIS offers both a GUI and command-line syntax. Commonly used core capabilities of GRASS GIS are listed below.

- Raster analysis: Automatic raster line and area to vector conversion; buffering of line structures; cell and profile data query; color table modifications; conversion to vector and point data format; correlation or covariance analysis; expert system analysis; map algebra; interpolation for missing values; neighborhood matrix analysis; raster overlay with or without weight; reclassification of cell labels; resampling resolution; rescaling of cell values; statistical cell analysis; surface generation from vector lines.

- 3D-raster analysis: 3D data import and export; 3D masks; 3D map algebra; 3D interpolation; 3D visualization; interface to Paraview and POVray visualization tools.

- Vector analysis: Contour generation from raster surfaces; conversion to raster and point data format; digitizing with mouse; reclassification of vector labels; superpositioning of vectors.

- Point data analysis: Delaunay triangulation; surface interpolation from spot heights; Thiessen polygons; topographic analysis (curvature; slope; aspect); LiDAR.

- Image processing: Support for aerial and UAV images; satellite data (optical, radar, thermal); canonical component analysis; color composite generation; edge detection; frequency filtering; histogram stretching; IHS transformation to RGB; image rectification; orthophoto rectification; principal component analysis; radiometric corrections; resampling; resolution enhancement (with RGB/IHS); RGB to IHS transformation; texture-oriented classification; shape detection; supervised classification; unsupervised classification.

- DTM-analysis: contour generation; cost/path analysis; slope/aspect analysis; surface generation from spot heights or contours.

- Geocoding: Geocoding of raster and vector maps including (LiDAR) point clouds.

- Visualization: 3D surfaces with 3D query; color assignments; histogram presentation; map overlay; point data maps; raster maps; vector maps.

- Map creation: Image maps; postscript maps; HTML maps.

- SQL-support: Database interfaces (DBF, SQLite, PostgreSQL, mySQL, ODBC).
- Geostatistics: Interface to R (a statistical analysis environment).
- Temporal framework: Support for time series analysis to manage, process, and analyze spatial–temporal environmental data.
- Others: Erosion modeling; landscape structure analysis; solution transport; watershed analysis, and so on.
- *Quantum GIS (QGIS)*: QGIS is a project of the Open Source Geospatial Foundation (OSGeo), a team of volunteers and organizations world-wide. Licensed under the GNU General Public License, QGIS is a free open-source GIS software package, including QGIS Desktop, QGIS Browser, QGIS Server, and QGIS Web Client. It runs on Linux, Unix, Mac OSX, Windows, and Android and supports numerous vector, raster, and database formats and functionalities. Its current release is QGIS 2.12. QGIS offers a variety of GIS functionalities provided by core features and plugins. Commonly used features and plugins are discussed in more detail here.
  - Viewing data: Users can view and overlay vector and raster data-sets in different formats and projections without the hassle of converting them into a common format or projection. This is a very convenient and powerful GIS functionality. Supported data formats include ESRI shapefile, MapInfo, SDTS, GML, GRASS raster and vector data from GRASS databases, GeoTIFF, ERDAS IMG, ArcInfo ASCII GRID, JPEG, PNG, spatially enabled tables and views using PostGIS, SpatiaLite and MS SQL Spatial, Oracle Spatial, online spatial data served as OGC Web Services, including WMS, WMTS, WCS, WFS, and WFS-T, and others.
  - Exploring data and composing maps: Users can compose maps and interactively explore spatial data with QGIS's easy-to-use GUI. Tools available in the GUI include QGIS browser, on-the-fly re-projection, database manager, map composer, overview panel, spatial bookmarks, annotation tools, identify/select features, edit/view/search attributes, data-defined feature labeling, data-defined vector and raster symbol tools, atlas map composition with graticule layers, north arrow/scale bar/copyright label for maps, support for saving and restoring projects, and so on.
  - Creating, editing, managing, and exporting data: Users can create, edit, manage, and export vector and raster data layers in a variety of data formats, as discussed above. Available tools include digitizing tools for OGR-supported formats and GRASS vector layers; creating and editing ESRI Shapefiles and GRASS vector layers; Georeferencer plugin to geocode images; GPS tools

for importing and exporting GPX format, converting other GPS formats to GPX, and uploading/downloading directly to/from a GPS unit; visualizing and editing OpenStreetMap data; creating spatial database tables from shapefiles with DB Manager plugin; managing vector attribute tables; saving screenshots as georeferenced images; DXF-Export tool; and plugins to perform CAD-like functions.

- Analyzing data: QGIS offers various vector analysis, sampling, geoprocessing, geometry, and database management tools. With the Processing Plugin, QGIS provides a powerful geospatial analysis framework to call native and third-party algorithms from QGIS, such as GDAL, SAGA, GRASS, fTools, and more. In addition, users can utilize the integrated GRASS tools to have access to the complete GRASS functionality of more than 400 modules to perform data analysis and many other GIS tasks.

- Publishing maps on the Internet: QGIS can be used as a WMS, WMTS, WMS-C or WFS, and WFS-T client, and as a WMS, WCS, or WFS server. Users can also publish data on the Internet using a webserver with UMN MapServer or GeoServer installed.

- QGIS plugins: QGIS functionality can be extended, enhanced, and customized through plugin architecture and libraries that can be used to create plugins. Users can create new applications and plugins with C++ or Python programming languages. Currently available QGIS Core plugins are listed below:

  - Coordinate capture: Capture mouse coordinates in different CRSs
  - DB manager: Exchange, edit, and view layers and tables; execute SQL queries
  - Dxf2Shp converter: Convert DXF files to shapefiles
  - eVIS: Visualize events
  - fTools: Analyze and manage vector data
  - GDALTools: Integrate GDAL tools into QGIS
  - Georeferencer GDAL: Add projection information to raster datasets using GDAL
  - GPS tools: Load, import, and export GPS data
  - GRASS: Integrate GRASS GIS
  - Heatmap: Generate raster heatmaps from point data
  - Interpolation plugin: Interpolate based on vertices of a vector layer
  - Metasearch Catalogue Client

- Offline editing: Allow offline editing and synchronizing with databases
- Oracle Spatial GeoRaster
- Processing (formerly SEXTANTE)
- Raster terrain analysis: Analyze raster-based terrain
- Road graph plugin: Analyze a shortest-path network
- Spatial Query Plugin
- SPIT: Import shapefiles to PostgreSQL/PostGIS
- Topology checker: Find topological errors in vector layers
- Zonal statistics plugin: Calculate count, sum, and mean of a raster for each polygon of a vector layer

- *gvSIG*: This GIS software name might look a little strange for English speakers. The first part, "gv," seems to have something to do with "government." But, what is "SIG"? Actually, in Spanish, Geographic Information System (GIS) is "Sistema de Información Geográfico (SIG)." So, its name indicates that it is a GIS software package for government(s). gvSIG was initiated by the Autonomous Valencian Government of Spain, and collaboratively developed by several companies, research organizations, and universities, including Prodevelop, Iver Tecnologías de la Information, Lógica Extrema, Open Sistemas, Universidad Jaume I, Universidad Politécnica de Valencia, Universidad de Valencia, Universidad Politécnica de Madrid, and so on. gvSIG is now a free open-source GIS package under the gvSIG Association, which have developed and released a family of GIS software products, including gvSIG Desktop, gvSIG Mobile, gvSIG Roads, gvSIG Educa, gvCity, gvNIX, and i3Geo. They are described briefly here.

  - gvSIG Desktop: Licensed under the GNU General Public License (GPL), gvSIG Desktop is free open-source software, which is free for use, distribution, study, and improvement. gvSIG Desktop offers a variety of features and tools to process, edit, analyze, and manage geographic data. It supports vector and raster files, databases, and remote services. gvSIG Desktop can be extended, improved, and customized to suit any special needs.

  - gvSIG Mobile: gvSIG Mobile is a GIS software product with a variety of tools to work with geographic information from mobile devices, running on Windows Mobile. It is a useful GIS software application for field data collection and updating. gvSIG Mobile supports vector, raster, and remote services. Also licensed under the GNU/GPL, gvSIG Mobile is free open-source software for use, distribution, study, and improvement.

- gvSIG Roads: This gvGIG software product serves the needs for end-to-end management of road infrastructures, such as task and job orders, guard tours, incidences, measurements, and certificates. It allows the maintenance of consistent working procedures, documentation, coordination with contractors, and access information. It is designed to be an easily extensible solution for inventory tasks, road safety, and expropriations, as well as being adaptable to the specific needs of each company. gvSIG Roads is free open-source software, under GNU General Public License.

- gvSIG Educa: This is a useful tool for educators in teaching geography and any other subjects with territorial components, such as history, economics, natural science, sociology, and so on. It helps students to better analyze and understand landscape through interactivity between students and information, and by adding spatial components to the study subjects. It also facilitates understanding of basic aspects by using visual tools such as thematic maps, and sharing knowledge and exercises between teachers and students. gvSIG Educa is free open-source software, under GNU General Public License.

- gvNIX: This is an open-source tool for quick web application development. It is a distribution of Spring Roo that provides the set of Spring Roo tools plus a suite of features that increase development productivity and improve the user experience by integrating frameworks like jQuery, Bootstrap 3, Leaflet, DataTables, Dandelion DataTables, and others. With tools like Hibernate Spatial and Leaflet, gvNIX can make it easy to develop web applications for management and visualization of geographic information. gvNIX automatically generates pages to display, list, search, create, and edit alphanumeric and geospatial data. It can also automatically generate independent geo-portals and/or application-integrated geo-portals.

- i3Geo: It is worth noting that i3Geo was developed by the Ministry of the Environment of Brazil and it is part of the Brazilian Public Software Portal. i3Geo integrates several open-source applications into a single development platform for generating interactive web maps. It is customizable and adaptable to meet different needs. It is also adaptable to different application programming interfaces (API), such as OpenLayers, Google Maps, and Google Earth.

- gvCity: This is currently under development, with limited information available.

- *Geostatistical R package (or RGeostats)*: Before RGeostats, the Centre de Géostatistique of the Ecole des Mines de Paris developed a few commercial libraries and software products. They include the GEOSLIB (a geostatistical library in FORTRAN), BLUEPACK (a geostatistical package for mining and oil industries), SIMPACK (for geostatistical stochastic simulations), HERESIM (a package based on the Plurigaussian simulation technique), and ISATIS (a geostatistical toolbox). Although these commercial software applications are useful, they cannot be modified for special needs. Therefore, starting in the late 1990s, researchers started to develop free and open-source applications using R interpreted language, such as the GEFA package developed to forecast fish density by species and age using geostatistical algorithms. With its free and open-source license, GEFA was significantly enhanced by many R developers. Later, it was rewritten with C and C++ compiled languages for better performance. Inspired by the success of GEFA, and based on a library of geostatistical code written in C and C++ for Geoslib, the RGeoS package was created in 2001. It contains a set of R objects to manipulate data, parameters, and results. In 2014, RGeoS was renamed as RGeostats.

RGeostats contains various functions and tools for geostatistical analyses, simulations, and modeling. Together with another newly developed package (R21), Geostatistical R Package is a free open-source software package. However, it must be used for a noncommercial purpose according to its End-User License Agreement. It supports Windows, Linux, and Mac platforms. Its current version is RGeostats v10.1.7, released in July 2015.

- *MapWindow GIS*: This is a free open-source "programmable geographic information system," which offers functionalities and tools for mapping, data viewing, manipulation, analysis, modeling, and a GIS application programming interface (API). Using MapWindow's application programming interface, a user can build a specialized GIS application to meet the needs of his or her own projects/studies, or sell it as a software product.

MapWindow GIS started with the project of developing watershed modeling software at Utah State University by a group of students and professors in 1998. They released MapWindow 1.0 as a set of GIS function tools in 2001. In 2004, convinced that it was not feasible for them to develop MapWindow into a commercial software product to compete in the ESRI-dominated GIS market, Utah State University and the project's primary sponsor, Idaho National Engineering and Environmental Laboratory (INL), decided to release MapWindow 3.0 and its source code to the public domain. It became a free open-source GIS software application, which attracted lots of interest from many individuals and organizations, such as the U.S. Environmental Protection Agency (EPA). The EPA was interested in using MapWindow as a platform for their Better Assessment Science Integrating Point and Nonpoint Sources (BASINS) watershed analysis and modeling system. With financial support from the EPA and other government and private organizations, MapWindow has been further developed and improved into a robust GIS software package. There are three desktop applications and two programming tools, which are discussed briefly here.

- MapWindow 4.x desktop: This is their major end-user desktop application. It is a free extensible GIS, which can be used as an open-source stand-alone desktop GIS application, as a data viewer to be bundled with datasets and distributed to others, developed as custom spatial data analysis tools, or embedded into other proprietary software. MapWindow Desktop supports a variety of data formats, including ESRI Shapefiles, ASCII and binary grids, GeoTIFF, and others. It contains functions of data visualization, data table and shapefile editing, data converting, and so on. It can be customized by editing the MapWindow Configuration File and/or a MapWindow Project File. They are XML format files, which can be easily edited using Notepad or an XML editor.

- MapWindow 6 alpha: This is a simple desktop GIS viewer application built for Microsoft Windows, based on the DotSpatial Library. The product is currently under development at Idaho State University together with our collaborators. Extensions for MapWindow 6 can be developed in C# or VB.NET by using the DotSpatial Library.

- HydroDesktop: As its name indicates, HydroDesktop is a hydrologic and climate analysis and modeling software tool, which was developed on DotSpatial. Hydrologic analysis and modeling are conducted in the OpenMI environment. The development was/ is led by Dr. Dan Ames and supported by the grants OCI-1148453, OCI-1148090, EAR-0622374, EPS-0814387, and EPS-0919514 from the National Science Foundation. Extensions can be developed for more advanced analysis of downloaded data, including integration with R and OpenMI. Data from HydroDesktop can also be exported for use in other applications.

- MapWinGIS ActiveX Control: This is the MapWindow Desktop application's core programmer library, which provides GIS and mapping functionalities. It is a C++-based GIS programming ActiveX Control and API. It can be embedded into a Windows Form in Visual Basic, C#, Delphi, and other languages that support ActiveX. Once embedded, it brings mapping and other GIS functionalities to the hosting software application. MapWinGIS. ocx is free and open source under the MPL 1.1 license, which allows it to be used freely in commercial and noncommercial applications.

- DotSpatial GIS Programmer Library: This is a GIS library written for .NET 4. It brings spatial data, analysis, and mapping functionalities into .NET applications. Its major functionalities are as follows:

  - Displaying a map in a .NET Windows Forms or Web application
  - Opening shapefiles, grids, rasters, and images
  - Rendering symbology and labels
  - Re-projecting on the fly
  - Manipulating and displaying attribute data
  - Enabling scientific analysis

- Reading GPS data

- *OpenJUMP GIS*: This is an open-source free (under the GUN GPL license) GIS software product based on JAVA Unified Mapping Platform (JUMP) GIS, developed by Vivid Solutions, based in Canada, using the JAVA programming language. It was a project for the British Columbia Ministry of Sustainable Resource Management to perform automated matching of roads and rivers from different digital maps into an integrated single geospatial dataset. OpenJUMP is the result of a larger umbrella project, the JUMP Pilot Project, which consists of English, French, and German branches, and other similar-minded organizations and individuals worldwide. OpenJUMP is mainly a vector GIS software package, with some image processing capabilities. It works on Windows, Linux, and MacOSX platforms. It can read Geography Markup Language (GML), SHP (ESRI shapefile), DXF, JML, MIF, TIFF, JPG, MrSID, and ECW, as well as data from PostGIS, ArcSDE, Oracle, and MySQL. It can export data to GML, SHP, DXF, PostGIS, JML, and JPG formats. Its current version is OpenJUMP 1.4. Its functions can be extended or enhanced by writing plugin(s), also called extensions, for special purposes beyond its core functions. The currently available plugins for OpenJUMP are

  - Editing plugins: Including ISA 2.5 tools for OpenJUMP deriving from SkyJUMP; CAD extension; measurement tools; set attributes extension
  - Raster image plugins: Including image plugin from Project PIROL; TANATO plugin from SIGLE; other raster plugins requiring Java Advanced Imaging (JAI) library
  - Print plugins: Including Basic Print PlugIn supporting multiple sheets by Geoff; Print Layout Plugin by Intevation

- Database plugins: The available database plugins are
  - DB query: enabling queries to retrieve and display geographic feature sets from PostGIS, MySQL, SpatialLite, and Oracle databases
  - PostGIS database connector
  - PostGIS plugin by Erwan Brocher
  - iSQLViewer: SQL front end by Erwan Brocher
  - ArcSDE database reader
  - Oracle database reader
  - SpatiaLite database reader
  - pgRouting network analysis with pgRouting and PostgreSQL
  - MonetDB datastore plugin
- Other data formats: Including DXF Driver (reading/writing AutoCAD dxf files); MIF/MID Driver (reading/writing mif-mid files); GeoConcept Driver (reading/writing GeoConcept files); Excel Driver (reading Microsoft Excel files); CSV Driver (reading/writing csv format files); OpenStreetMap osm file Driver (reading OpenStreetMap xml files)
- Internet data sources and web processing: Including WFS PlugIn/WFS-T PlugIn for OGC WFS Standards 1.0 and 1.1 by Lat/Lon; WPS PlugIn for OGC Web Processing Standard 1.0 by 52 North
- Spatial analysis and editing plugins: The available plugins are
  - Map Generalization Toolbox.
  - Sextante: A set of free geospatial analysis tools developed by the Junta de Extremadura (Spain). Its main aim is to create a platform for the development of algorithms for raster and vector analysis on an independent java library that can work on different GISs (such as geotools, Gvsig, Openjump, and others).
  - Clean and explode: Editing/improving ArcGIS-generated contour lines.
  - Landscape ecology: Extracting edges, core area, patches, corridors, branches, and shortcut areas with buffer operations.
  - Quality Assurance Plugin.
  - Road Matcher Plugin.
  - Graph Toolbox Plugin.
  - Skeletonizer Plugin.
  - GeOxygene Plugin: Developed for the GeOygene analysis and computation platform.

- Home Range Analysis Toolbox (HoRAE): This can be used for animal movement analysis based on GPS collar data.
- MorphAL: This can be used for morphological analysis of vector data.
- Multi Depot Vehicle Routing.
- Matching PlugIn: This can be used to find matching features in one layer or between two layers, based on geometric and/or semantic criteria. It has advanced options to do fuzzy matching, such as geometry matching based on Hausdorff distance and string matching based on Levenshtein distance.
- Space Syntax Analysis (SSA) Plugin: Performing SSA for convex spaces (polygon networks) with functions available in certain FOSS4GIS. SSA Plugins calculate the basic space syntax measures including connectivity, total depth, mean depth, global integration, local depth, local integration, and control values for each basic spatial unit involved in a spatial configuration, and intelligibility value for the whole of the configuration.
- Concave Hull Plugin.

- Raster analysis: Including OpenKlem, which is an extension of OpenJUMP for hydrological analysis. It includes a module for simulating flood hydrographs (Kinematik Local Excess Model, KLEM).
- Coordinate reference systems/coordinate Transformation: including CTS extension, which allows users to assign a coordinate reference system to vector layers and re-project data layers.
- Security for Geodata.
- Programming and scripting support: including Beanshell Editor, which is a text editor used to write, launch, and save Beanshell scripts; Groovy.
- GPS plugins: including Live GPS extension; GPX import plugin; nmeaRaw.jar Plugin; NMEA converter script.
- Styling: including VertexSymbols Plugin, which adds vector or raster symbols to vertex; Jump Fill Pattern Plugin, which allows users to add their own fill patterns to an Openjump project from Cadplan.
- Charts/plots: a set of plugins that provide attribute classification tools and charts (scatter plot, bar plot, histogram).
- Selection Tools Package
- SVG image export plugin
- Proportional symbols plugin

- The Looks Extension: applying the aesthetically pleasing JGoodies Looks look-and-feel to Openjump.
- ISA Tools Package: containing several useful tools especially for geometry editing, displaying MrSid raster images, and connecting to the ESRI ArcSDE geodatabase.

- *uDig*: UDig is defined as a user-friendly, desktop-located, Internet-oriented, and GIS-ready open-source software application framework, distributed under the Eclipse Public License. It runs as a thick client on Windows, Mac OS/X, and Linux. Based on Eclipse Rich Client Platform (RCP) technology, UDig was developed by a community led by Refractions Research, a geospatial software company in Canada. Its programming language is JAVA. Its current version is Release 1.5.0.RC1.

  uDig can be used as a stand-alone application, or as a plugin in an existing RCP application. It has capabilities to view, edit, and print files, databases (Oracle, SDE, PostGIS), and Web Feature Servers (WFS). Its functionalities can be extended with RCP plugins. The Eclipse platform is designed to serve as an open tools platform. One of the advantages of the Eclipse platform is that its inherent extensibility allows developers to build not only a closed-form product, but also an open-ended platform in their own domain for further development and enhancement. RCP refers to a minimal set of plugins needed to build a rich client application. Below is a list of the major components and plugins included in the Eclipse RCP.

  - Eclipse runtime: Building on top of the OSGi framework, Eclipse runtime provides the foundational support for plugins, extension points, and extensions. Its plugins include the org.eclipse. core.runtime, org.eclipse.osgi, and org.eclipse.osgi services.
  - The Standard Widget Toolkit (SWT): The SWT provides efficient and portable access to the user interface facilities of the operating systems on which it is implemented. Its plugins include the org. eclipse.swt and platform-specific fragments.
  - JFace: Building on top of the Standard Widget Toolkit (SWT), JFace is a user interface (UI) framework for handling many common UI programming tasks. It has one plugin: org.eclipse.jface.
  - Workbench: Building on top of the Eclipse Runtime, SWT, and JFace, Workbench provides a highly scalable, open-ended, multi-window environment for managing views, editors, perspectives, actions, wizards, preference pages, and so on. Its plugins includes org.eclipse.ui, org.eclipse.ui.workbench, org.eclipse.core.expressions, org.eclipse.core.commands, and org.eclipse.help.
  - Update manager: Allows users to discover and install updated versions of products and extensions. Its plugins include org.

eclipse.update.configurator, org.eclipse.update.core, org.eclipse. update.scheduler, org.eclipse.update.ui, and other platform-specific fragments.

- Forms: A flat-look control library and multipage editor framework. It has one plugin: org.eclipse.ui.forms.

- Help: A web-app-based Help UI, with support for dynamic content. Its plugins include org.apache.lucene, org.eclipse.help. appserver, org.eclipse.help.base, org.eclipse.help.ui, org.eclipse. help.webapp, and org.eclipse.tomcat.

- Text: A framework for building high-function text editors. Its plugins include org.eclipse.text, org.eclipse.jface.text, and org. eclipse.workbench.texteditor.

- Resources: A workspace resource model, with managed projects, folders, and files. It has one plugin: org.eclipse.core.resources.

- Eclipse Modeling Framework (EMF) and Service Data Objects (SDO): EMF is a modeling framework and code-generation facility for building tools and other applications based on a structured data model. SDO is a framework that simplifies and unifies data application development in a service-oriented architecture (SOA).

- Graphical Editing Framework (GEF): A framework for building graphical editors, such as Draw2D, which is a vector graphics framework. Its plugins include org.eclipse.draw2d and org. eclipse.gef.

- Console: Extensible console view. It has one plugin: org.eclipse. ui.console.

- Cheat Sheets: A Cheat Sheet guides the user through a long-running, multistep task. It has one plugin: org.eclipse. ui.cheatsheets.

- *System for Automated Geoscientific Analyses (SAGA GIS)*: Licensed under the GNU General Public License, SAGA GIS is now a free open-source Software (FOSS), running under Windows and Linux operating systems. The original idea of the development of SAGA came from a few research projects worked by a small team in the late 1990s, at the Department of Physical Geography, University of Göttingen, Germany. One of their research topics was analyzing Digital Elevation Models (DEM) raster data to predict soil properties, terrain-controlled process dynamics, and climate parameters. Their development and implementation of some new methods for spatial analysis and modeling formed the basis for the development of SAGA GIS in 2001. SAGA is coded in C++ programming language, with an industrial standard object-oriented system design. In February 2004, the research and development team decided to publish SAGA v1.0 as a free open-source software package to share it with geoscientists worldwide and attract contributions from others to further develop and improve it. In order to establish a sustainable long-term development arrangement, the SAGA User Group Association, a nonprofit organization, was founded in early 2005 to coordinate development efforts, organize user group meetings, and perform other tasks. The center of the SAGA development team is now located at the Institute of Geography of the University of Hamburg, Germany.

SAGA functions are organized as modules in framework-independent module libraries. As Dynamic Link Libraries (DLL), module libraries are not independently running executables. They have to be accessed through SAGA's GUI or various scripting environments, such as shell scripts, Python, R, and so on. SAGA's GUI includes a menu, a toolbar, a status bars, and three workspaces, which are subwindows for the modules, data, and maps. The latest version, SAGA GIS 2.2, contains more than 300 modules for georeferencing/cartographic projections, grid interpolation, vector tools, image analysis, geostatistics, terrain analysis, and so on.

- *GeoServer*: Built on Geotools, which is an open-source Java GIS toolkit, GeoServer is a free open-source GIS software server, which allows users to publish, view, and edit geospatial data. GeoServer complies with Open Geospatial Consortium (OGC), Web Map Service (WMS), Web Coverage Service (WCS), and Web Feature Service (WFS) standards. GeoServer is a core component of the Geospatial Web or Geoweb. Similar to the famous World Wide Web (WWW), Geoweb is an interconnected computer network that allows users to publish, search for, and download spatial data. With GeoServer, users can generate maps quickly and easy, using OpenLayers, a free mapping library integrated into GeoServer. Maps can be exported in a variety of output formats. Map data layers can be edited, and shared so that other users can use the data in their websites and applications. GeoServer can display data on a variety of existing popular mapping applications, such as Google Maps, Google Earth, Yahoo Maps, NASA World Wind, Bing Maps, and Microsoft Virtual Earth. It can also connect with ESRI ArcGIS, a widely used commercial GIS software package. Its current release is GeoServer 2.8.1.

  GeoServer was one of a suite of tools developed in 2001 by the Open Planning Project (TOPP), a nonprofit technology organization based in New York to enable citizen involvement in government and urban planning by sharing spatial data. Since spatial data are mainly stored in ESRI shapefile and ArcSDE-enabled Oracle and other geodatabases, GeoTools was integrated into GeoServer to support these widely used spatial data formats. In the meantime,

PostGIS, a free and open spatial database, was developed by Refractions Research. New features were added to GeoServer so that it can connect to this free geodatabase. There are many extensions, also known as modules, available to add more capabilities to GeoServer. With them, GeoServer supports numerous data formats, such as Shapefiles, GeoTIFF, ECW, MrSID, JPEG2000, JPEG, GIF, SVG, PNG, KML, GML, PDF, GTOPO30, GeoRSS, GeoJSON, PostGIS, ArcSDE, Oracle Spatial, DB2, MySQL, and others.

- *PostgreSQL*: PostgreSQL is a free open-source object-relational database management system (ORDBMS), under a permissive free-software license. It is cross-platform and runs on a variety of operating systems, such as, Windows, UNIX, Linux, Mac OS, FreeBSD, Solaris, and others. PostgreSQL came from the Interactive Graphics Retrieval System (Ingres) database management system project at the University of California, Berkeley (UCB) in the early 1980s. Ingres was developed based on relational database management system (RDBMS) concepts. Later, the project was renamed as post-Ingres, and then to POSTGRES in the mid-1980s, because the new ORDBMS concepts were introduced to the project. It was continuously improved until the project was terminated in June 1994, and POSTGRES v.4.2 was released under a special free open software license, to allow other developers to use its code and further develop it. Later, support for SQL query language was added to the project, and therefore POSTGRES was renamed as PostgreSQL in 1996. PostgreSQL v6.0 was released as free open software in January 1997. Since then, the PostgreSQL Global Development Group was formed by a group of developers and volunteers around the world to maintain and further improve PostgreSQL. Its current version is PostgreSQL 9.5.

    PostgreSQL contains built-in support for three procedural languages, that is, plain SQL, PL/pgSQL, and C. With language extensions, PostgreSQL supports a variety of others procedural languages, such as Perl, Python, Java, JavaScript, R, Tcl, and so on. It is compliant to Atomicity, Consistency, Isolation, and Durability (ACID) to ensure that its database transactions are processed reliably. Similar to other object-relational database management systems, PostgreSQL has support for foreign keys, joins, views, triggers, and stored procedures. As an enterprise database system, PostgreSQL is scalable for both the quantity of data it can manage and the number of concurrent users it can serve. Using the built-in multiversion concurrency control (MVCC) feature, PostgreSQL eliminates read locks. Below is a list of PostgreSQL's technical limits and their corresponding values.

    - Maximum database size: Unlimited
    - Maximum table size: 32 TB

- Maximum row size: 1.6 TB
- Maximum field size: 1 GB
- Maximum rows per table: Unlimited
- Maximum columns per table: 250–1600, depending on column types
- Maximum indexes per table: Unlimited

PostgreSQL has many programming interfaces, including C/C++, Java, .Net, Perl, Python, Ruby, Tcl, ODBC, and so on. In addition to its two built-in programming interfaces, that is, libpq (a C application interface) and ECPG (an embedded C system), PostgreSQL supports many external programming interfaces too. Listed below are some of the supported external programming interfaces.

- libpqxx: C++ interface
- PostgresDAC: PostgresDAC
- DBD::Pg: Perl DBI driver
- JDBC: JDBC interface
- Lua: Lua interface
- Npgsql: .NET data provider
- ST-Links SpatialKit: Link Tool to ArcGIS
- PostgreSQL.jl: Julia interface
- node-postgres: Node.js interface
- pgoledb: OLEDB interface
- psqlODBC: ODBC interface
- psycopg2: Python interface
- pgtclng: Tcl interface
- pyODBC: Python library
- php5-pgsql: PHP driver based on libpq
- postmodern: Lisp interface

Administering tools for PostgreSQL include psql (a command-line administration tool), pgAdmin (a graphical user interface administration tool), phpPgAdmin (a web-based administration tool), PostgreSQL Studio, TeamPostgreSQL, pgFouine (PostgreSQL log analyzer), and LibreOffice/OpenOffice.org Base. There are many extensions (also known as add-ons) to extend and/or enhance the features and functionalities of PostgreSQL. A few example extensions are listed and briefly explained here. Please note that some extensions/add-ons are not free open-source software.

- OpenFTS (open-source full text search engine): providing online indexing of data and relevance ranking for database searching
- GiST (generalized search tree) indexing: a system of sorting and searching algorithms
- MADlib: providing mathematical, statistical, and machine-learning methods for structured and unstructured data
- MySQL migration wizard
- Performance Wizard
- PostGIS: providing support for spatial objects, similar to ESRI's SDE or Oracle's Spatial
- pgRouting: extending PostGIS to provide geospatial routing functionality
- Postgres Enterprise Manager: providing remote monitoring, management, reporting, capacity planning, and tuning
- ST-Links SpatialKit: for directly connecting to spatial databases

  PostgreSQL supports a large number of data types, such as Boolean, Arbitrary precision numeric, Character (text, varchar, char), Binary, Date/time, Money, Enum, Bit strings, Text search type, Composite, HStore, Arrays, Geometric primitives, IPv4/IPv6 addresses, CIDR blocks and MAC addresses, XML supporting XPath queries, UUID, JSON, and so on.

- *SQLite*: SQLite is an RDBMS engine embedded into other end programs. Unlike other conventional database management systems, such as Oracle, SQL Server, MySQL, DB2, PostgreSQL, Informix, SAP Sybase Adaptive Server Enterprise, SAP Sybase IQ, and Teradata, SQLite is not a stand-alone DBMS. The host application program accesses SQLite's functionalities through function calls. Also, since SQLite is embedded inside the application, internal function calls are more efficient than interprocess communication between an application and an external database. Development of SQLite was started in early 2000 for a project that needed a program to be operated without the support of a separate database management system, that is, an application containing a database system inside it. Later on, it became free open-source software. It is now maintained and under further development by the SQLite Consortium, with the following members: Bentley, Bloomberg, Navigation Data Standard (NDS), Expensify, Mozilla, and others. SQLite runs on many major operating systems, such as Windows, Linux, webOS, Mac OS, Android, Solaris, Tizen, Maemo, BlackBerry OS, Symbian OS, NetBSD, OpenBSD, FreeBSD, and others. It supports a variety of programming languages. A few key features of SQLite are listed here.

- ACID (Atomic, consistent, isolated, and durable) transactions.
- Zero-configuration: As an embedded DBMS, no setup or configuration is needed.
- Full SQL implementation.
- A complete database stored in a single cross-platform disk file.
- Supporting terabyte-sized databases and gigabyte-sized strings and blobs.
- Small code footprint.
- Simple and easy-to-use API.
- Self-contained: without any external dependency.
- Cross-platform.
- A stand-alone command-line interface (CLI) client available for database administration.
- Providing vector geodatabase functionality through SpatiaLite extension, similar to PostGIS (for PostgreSQL), Oracle Spatial, and ESRI ArcSDE.

- *Integrated Land and Water Information System (ILWIS)*: The Institute for Aerospace Survey and Earth Sciences (ITC) of the University of Twentein in the Netherlands developed a GIS software application (named as DGIS) for land use planning and watershed management studies. It was funded by a grant from the Dutch Ministry of Foreign Affairs to ITC, in late 1984. DGIS was one of the earliest fully integrated raster and vector GIS software products in the world, with tools for remote sensing data. It was primarily used and tested within ITC. It was further developed into the Integrated Land and Water Information System, and remote sensing program (ILWIS 1.0), running in DOS operating system, in late 1988, and distributed as a commercial GIS software product in 1990. When Microsoft Windows operating system became popular and dominant in the PC world in the 1990s, ILWIS was re-engineered for the Windows operating system. ILWIS 2.0 for Windows was released in late 1996, and ILWIS 3.0 was distributed in mid-2001. ILWIS 3.2 was released as shareware (1 month trial offer) in January 2004. ILWIS became an open-source software application under GPL in July 2007. Since then, it has been improved, updated, and distributed by 52°North, an open international network of partners from research, industry, and public administration. Since the office of this organization is located in the city of Münster, Germany, where the 52°N latitude line passes, it was named 52°North.

  Its current version is ILWIS 3.8. Although ILWIS only runs on Microsoft Windows 98 and higher, it can run on a Mac or Linux, using WINE. Major capabilities of ILWIS include data viewing, digitizing, editing, processing remote sensing and other raster images, data analysis, modeling, 3D visualization, mapping, and so on. Its key features are listed here.

- Integrated raster and vector design
- Import and export modules of widely used data formats
- On-screen and tablet digitizing
- Comprehensive set of image processing tools
- Orthophoto, image georeferencing, transformation, and mosaicking
- Advanced modeling and spatial data analysis
- 3D visualization with interactive editing for optimal view findings
- Rich projection and coordinate system library
- Geostatistical analyses, with Kriging for improved interpolation
- Production and visualization of stereo image pairs
- Spatial Multiple Criteria Evaluation
- Animation framework (with optional 3D)
- Auto resampling of different spatial geometries
- Surface energy balances
- Set of operations on DEMs/DTMs and hydrological processing
- Hydrological modeling
- Web Processing Service (WPS)
- Web Map Service Interface Standard (WMS).

WMS provides an HTTP interface for requesting georegistered images from distributed geospatial databases. A WMS request is able to define the geographic data layers, as well as area of interest. The response to the request is one or more georegistered images in JPEG, PNG, or other formats, which can be displayed in a browser application.

Major Commercial GIS Software Vendors and Products

- *ESRI GIS software products*: Currently, the ESRI is the dominant GIS software vendor. It entered the commercial GIS software market in 1982 with its command-line-based GIS software product, ARC/INFO, which attracted a lot of attention in the GIS world. Back then, although it was a useful GIS software application that was capable of data processing (both raster and vector), editing, analyzing, modeling, and map production, it was difficult to understand, memorize, and use its commands and associated parameters. The ESRI took note of these shortcomings and the complaints from users, and significantly improved it with the release of ARC/INFO v. 4.0 and the ARC Macro Language (AML) in 1986. Commands and their parameters could be written into AML programs and executed automatically. Simple GUIs, such as menus and forms, could also be created with AML. With the software improvements and the AML programming language, ARC/INFO became much easier to use, more efficient, and productive. It started to gain popularity worldwide. However, it was the launch of the GUI-based desktop ArcView in 1991 that further differentiate ESRI from other GIS software vendors and established its dominant status. The ESRI's GIS software products and their data formats (e.g., shapefile and geodatabase) have been referred to as industry standards and used by governments, industries, and organizations worldwide. Their current GIS software packages include ArcGIS 10.x for Desktop, ArcGIS 10.x for Server, ArcPad 10.x (a mobile system with GIS and GPS capabilities), ArcReader (a free desktop application for viewing GIS data and printing simple maps), ArcGIS Online, and so on. It is worth noting that ArcGIS Online is an account-based web GIS service controlled by the ESRI, which allows users to use, create, share maps and data, add items, and publish web layers.

- *Intergraph*: Intergraph was founded as M&S Computing in 1969, and renamed as Intergraph Corporation in 1980. It was acquired and restructured by Hexagon in 2010. Its geospatial technology is now under the Hexagon Geospatial division. Its current GIS and remote sensing related software packages include GeoMedia GIS, GeoMedia Add-Ons, GeoMedia Smart Client, ERDAS IMAGINE, ERDAS IMAGINE Add-Ons, ERDAS ER Mapper, LiDAR, and so on.

- *Smallworld*: This was originally developed in 1989 by the Smallworld Company in Cambridge, England, and was acquired in 2000 by GE Digital Energy, a division of General Electric. It is a GIS software product mainly designed for power and other utilities, telecommunications, and related industries.

- *Bentley Systems*: Bentley Systems was founded in 1984. It is better known for its flagship MicroStation, which is a CADD software package mainly for the design, construction, and operation of infrastructure. Its desktop GIS software is Bentley Map, developed mainly for infrastructure-related uses. Its Enterprise GIS packages include Bentley Geospatial Server, ProjectWise Connector for ArcGIS, ProjectWise Connector for Oracle, and ProjectWise Geospatial Management. It also offers other GIS packages, such as Bentley Geo Web Publisher, Bentley Map Mobile, Bentley Utilities Designer, MineCycle Survey, and so on. Their CADD and GIS software products are widely used by government agencies, and the utility, transportation, and communications industries.

- *Autodesk*: Autodesk was founded in 1982, in San Rafael, California. It is better known for its AutoCAD software package, which is widely used in architecture, engineering, construction, manufacturing, and other industries. Its GIS-related software packages include AutoCAD Map 3D, Infrastructure Map Server, InfraWorks 360, and so on.

- *MapInfo*: MapInfo was founded in 1986 by a group of students at Rensselaer Polytechnic Institute in Albany, New York, to develop relatively cheap mapping tools for personal computers competing with ESRI's expensive GIS software product(s) for workstations. In 2007, Pitney Bowes acquired MapInfo Corporation. Their GIS software package is the desktop MapInfo Pro with a few plugins, including MapInfo Drivetime for route analysis, MapInfo Engage 3D for 3-dimensional and statistical analysis, and MapInfo MapMarker for Geocoding. Its current version is MapInfo Professional v12.5.

- *Geosoft*: Geosoft was established in 1986 by two Canadian earth scientists, Ian MacLeod and Colin Reeves, to develop software products originally with FORTRAN for geophysicists and geologists. They also developed GIS extension(s) and plugins mainly for geologic and geochemical data analysis and mapping within ESRI's desktop ArcGIS package, such as Target for ArcGIS (a geology mapping extension), Geochemistry for ArcGIS, and others.

GIS Technology Widely Used in Environmental and Earth Sciences

Some environmental problems can be large-scale, long-term, and complicated. A sufficient amount of data are required to understand the problems, evaluate their current and possible future impacts on human health, environment, and ecosystems, and find the most effective strategies to mitigate and/or solve them. GIS technology offers powerful tools for collecting, processing, editing, storing, managing, sharing, analyzing, modeling, and visualizing large volumes of data to help scientists, decision-makers, stakeholders, and the general public to better understand environmental problems, find solutions, and protect the environment.

In the last 10 years or so, the use of GIS technology in environmental and earth sciences has grown phenomenally because GIS technology has become more mature, reliable, and powerful after more than 50 years of development and enhancement. There are a wide range of free open-source and commercial GIS software products/packages available for virtually all industries, including environmental and earth sciences. Also, the extraordinary improvements in the computing industry in recent years, especially in data processing, storage, and distribution, have fueled the growth of GIS because spatial datasets usually have large file sizes and therefore demand intensive computing resources and huge system memories and storage. The Internet and cloud computing technologies make data distribution unprecedentedly faster, cheaper, and easier. Realizing the advantages of GIS technology, more people have started to learn and use GIS. Spatial data and metadata standards are being developed and adopted by more and more governments and industries, which makes it easier to share GIS data across industries and government agencies.

Nowadays, GIS technology is widely used in virtually all aspects and areas of environmental industry, such as environmental disaster assessment, responses, and monitoring (e.g., large-scale oil spills on land and in water, nuclear material leaking, sewer spills, fires, and flooding involving hazardous materials); global, national, and/or regional scale hazardous sites inventorying, characterization, and monitoring; environmental impact studies for construction, development, recreation, and tourism projects; environmental permitting, compliance auditing and reporting; environmentally impacted site/area investigation, mitigation, and/or remediation; environmental and resource conservation; and many more. In earth science, GIS is extensively used in geologic mapping/survey; geophysical survey; hydrological studies; plate tectonics, volcano and earthquake research; geotechnical investigations; topographic analysis; slope stability studies; soil characterization, mapping and conservation; erosion control; mineral, oil, and gas explorations; and so on. Environmental science and earth science are closely related disciplines. Many environmental projects have geologic, geophysical, and/or hydrological study components because data about the project site's geologic and hydrological conditions are essential in understanding environmental problems and designing strategies and approaches to solve them.

As a leading global environmental authority, the UN Environment Programme (UNEP) uses GIS technology in inventorying and monitoring global and regional hazardous waste sites; evaluating environmental impacts from various conflicts and disasters and exploring solutions; assessing global, regional, and national environmental conditions and trends; promoting sustainable development within the UN system; and inspiring, informing, and enabling nations and peoples to protect the environment while improving their quality of life.

The UNEP developed an online database containing a large number of geospatial datasets, including freshwater, population, forests, emissions, climate, disasters, health GDP, and so on. The datasets are used by the UNEP and its partners in the Global Environment Outlook (GEO) reports and integrated environment assessments of national, subregional, regional, and global scales. With the UNEP's online Environmental Data Explorer, these valuable datasets can be displayed on the fly as maps, graphs, and data

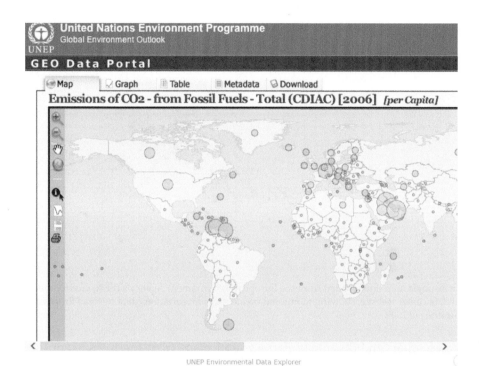

UNEP Environmental Data Explorer

FIGURE 1.14

A screenshot of the UNEP's online GIS map, illustrating the UNEP's global assessment of CO_2 emissions from fossil fuel per capita in 2006. The unit used in the calculation of the amount of the CO_2 emission is gigagram (1,000,000,000 grams). Each country or region is represented by a blue circle. The size of the circle is proportional to the CO_2 emissions from fossil fuel per capita in that country or region, ranging from 0.00002 gigagram (represented by the smallest blue circle) to 0.04621 gigagram (represented by the largest blue circle).

tables. From the online Environmental Data Explorer, users can explore and download the datasets in different formats. Figure 1.14 is an example online map showing UNEP's global assessment of CO_2 emissions from fossil fuel per capita in 2006.

The European data centers, managed and maintained by the European Environment Agency (EEA), European statistics (EUROSTAT), and the Joint Research Centre (JRC), contain a wide range of spatial datasets and GIS maps, such as European air pollution, waste sites, biodiversity, water, natural resources, forest, and climate change. The datasets and maps cover the member nations of Greece, Poland, Romania, Portugal, Spain, the United Kingdom, the Netherlands, Belgium, Germany, France, Czech Republic, Italy, Cyprus, Estonia, Latvia, Lithuania, Finland, Hungary, Bulgaria, Malta, Denmark, Iceland, Switzerland, Sweden, Austria, Luxembourg, Ireland, Liechtenstein, Slovakia, Norway, and Slovenia. Users can submit their datasets to the data centers following the data submission guidelines, and

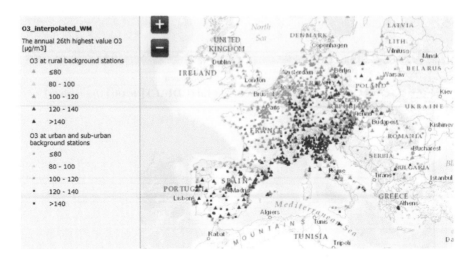

FIGURE 1.15

An example map downloaded from the European Environment Agency's (EEA) environmental data center website, showing monitored ozone (O_3) concentrations data in some European countries in 2009.

search, view, and download datasets and maps from the websites of the data centers. Figure 1.15 is an example map from the EEA's environmental data center website, showing ozone (O_3) concentration data in some European countries in 2009. The ozone data were reported from the monitoring stations in an air monitoring network. Ozone is an air pollutant, which is the primary constituent of smog at ground level. It is created by sunlight acting on mono-nitrogen oxides (NO), nitrogen dioxide (NO_2), and volatile organic compounds (VOCs) in the air, including gasoline vapors, chemical solvents, combustion products of fuels, and others.

National, regional, statewide/provincial, local governments, and other related agencies using GIS mapping, as well as data collecting, processing, managing, analyzing, and modeling tools to inventory environmentally impacted sites and hazardous waste producing facilities under their jurisdictions, track permits, audit compliance, investigate and document violations, evaluate and respond to environmental emergencies, initiate and monitor mitigation and/or remediation efforts, promote clean and sustainable production and development, inform and educate stakeholders and the general public about environmental issues and the best practices, and so on.

For example, the United States Environmental Protection Agency (USEPA) compiles and maintains spatial databases containing detailed information of the Superfund sites in the United States, designated under the Comprehensive Environmental Response, Compensation, and Liability Act (CERCLA) of 1980, enacted by Congress in Public Law 96-510. Superfund sites are severely polluted areas which require long-term efforts to clean up hazardous material contaminations. Per CERCLA, Superfund sites are also placed on the

National Priorities List (NPL). Following the guidelines of the NPL, EPA further characterizes and determines the sites which need further investigation for environmental remediation. Designed to protect public health and the environment, EPA's Superfund Program is responsible for cleaning up some of the nation's most contaminated land and responding to environmental emergencies, oil spills and natural disasters. Figure 1.16 is a GIS map showing the general locations of the Superfund sites in the United States. This GIS map was generated with the GIS shapefiles and geodatabases downloaded from EPA's Superfund Program website. As of February 27, 2014, there were 1322 Superfund sites on the National Priorities List, with additional 53 sites proposed for entry on the list. Three hundred and seventy-five Superfund sites have been cleaned up and removed from the list.

The U.S. Army Corps of Engineers (USACE) also supports the Environmental Protection Agency's (EPA) Superfund Program, because the CERCLA authorizes EPA to utilize the existing capabilities of other Federal agencies in meeting its objectives. Upon EPA request, USACE may assist EPA with remedial investigation and feasibility study (RI/FS), remedial design and remedial action, environmental impact statements, obtaining permits, legal determinations, obtaining real estate, and so on.

USACE is also responsible for environmental restoration of properties that were formerly owned by, leased to or otherwise possessed by the United States and under the jurisdiction of the Secretary of Defense. These types of properties are designated as Formerly Used Defense Sites (FUDS), which were used by the Department of Defense (DoD) for a variety of purposes, including training and supporting soldiers, airmen, sailors, and Marines, as well as to test new weapons and warfare capabilities, according to the information on the USACE's FUDS website. The U.S. Army Corps of Engineers executes the FUDS program on behalf of the Army and the Department of Defense. The FUDS is a huge and complex program with more than 10,000 properties identified for potential inclusion in it. The FUDS can range from less than an acre to hundreds of thousands of acres with potential hazardous, toxic and/or radioactive contaminants. They can be in industrial or residential areas as well as on federal, tribal or state properties. The U.S. Army Corps of Engineers uses GIS technology extensively in inventorying these environmentally impacted sites, collecting sufficient amount of data to identify FUDS eligible properties, investigating the origins and extents of contamination, assessing their conditions, evaluating their potential risks to human health, environment and ecosystems, addressing land transfer issues, tracking past and present property ownerships, documenting stakeholders, informing and educating the public, and so on. Similar to EPA's Superfund sites, environmental cleanup activities at DoD's FUDS properties must be conducted in accordance with the Comprehensive Environmental Response, Compensation and Liability Act too. Figure 1.17 is a FUDS locations GIS map generated from the GIS datasets downloaded from the USACE FUDS program website.

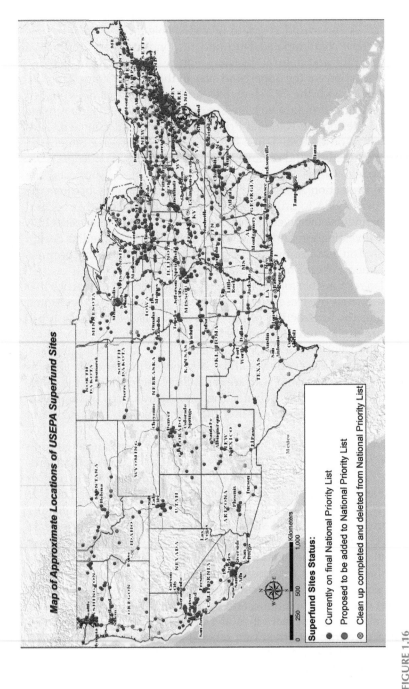

Map of Approximate Locations of USEPA Superfund Sites

Superfund Sites Status:

- Currently on final National Priority List
- Proposed to be added to National Priority List
- Clean up completed and deleted from National Priority List

FIGURE 1.16

A GIS map showing the general locations of the USEPA Superfund sites, created from the GIS data (i.e., shapefiles and geodatabases) downloaded from the United States Environmental Protection Agency's (USEPA) Superfund Program website. The red points represent the Superfund sites currently on the final National Priority List. The blue points represent the Superfund sites proposed to be added to the National Priority List. The green points represent the Superfund sites which have been cleaned up and deleted from the National Priority List.

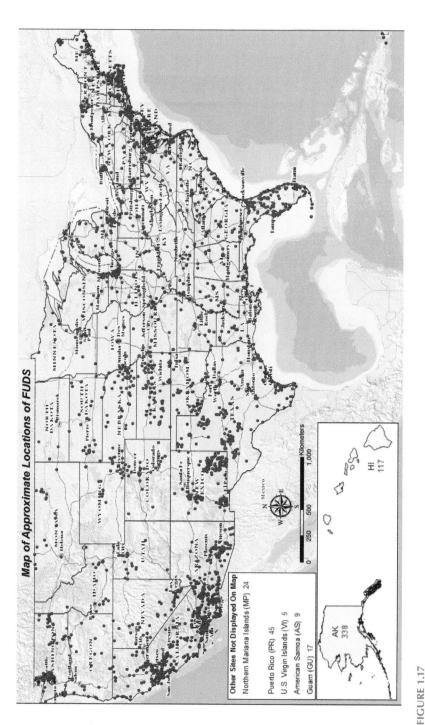

FIGURE 1.17

General FUDS locations map created from the GIS data layers (i.e., shapefiles) obtained from the U.S. Army Corps of Engineers (USACE) FUDS program. The red points on the main U.S. map represent the approximate locations of the FUDS. For the states of Alaska (AK) and Hawaii (HI), and some U.S. territories, only the total numbers of FUDS were shown or listed in the insets. There are 338 FUDS in Alaska (AK), 117 FUDS in Hawaii (HI), 24 FUDS in the Commonwealth of the Northern Mariana Islands (CNMI), 45 FUDS in Puerto Rico (PR), 5 FUDS in U.S. Virgin Islands (VI), 9 FUDS in American Samoa (AS), and 17 FUDS in Guam (GU).

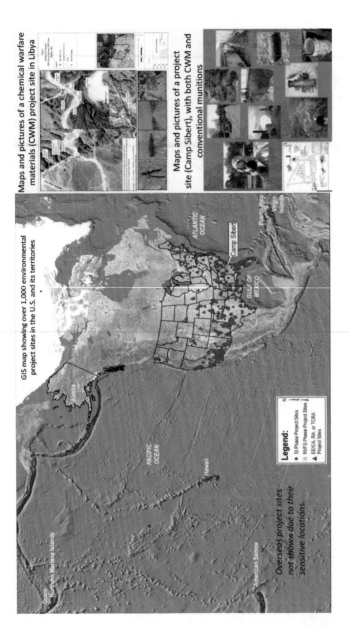

FIGURE 1.18

Locations of the over 1000 environmental projects sites in the United States and its territories, the author has supported with GIS technology. The red points represent the Site Inspections (SI) project sites. The green squares represent the Remedial Investigations and Feasibility Studies (RI/FS) project sites. The purple triangles represent the Engineering Evaluation and Cost Analysis (EE/CA), Preliminary Assessment (PA), and Time Critical Removal Action (TCRA) project sites. On the right side of this Figure, the top map displays a chemical warfare material (CWM) project site in Libya. The photos below the map show some unexploded ordnance (UXO), munitions constituents in the CWM project site. The photos in the bottom right of the figure show some chemical and conventional unexploded ordnance (UXO), munitions debris (MD), munitions constituents, and a Personal Protective Equipment (PPE), i.e., the white suite and a respiratory mask required for performing work in a hazardous site. The type or level of a PPE is determined by the conditions of the project site.

In the past 30 years, the author has worked on hundreds of environmental and geology projects, with more than one thousand project sites across the United States, its territories (Guam, Northern Mariana Islands, American Samoa, Saipan, Puerto Rico, and Virgin Islands), and other countries, such as China, Japan, Iraq, Libya, Kuwait, Saudi Arabia, Rwanda, Haiti, and so on. Figure 1.18 is a GIS map showing the environmental project sites in the United States and its territories. The international project sites are not shown in this map because the locations of some sites are highly sensitive either politically or militarily.

Some of these environmental projects sites are Formerly Used Defense Sites (FUDS), containing military munitions (MM) and munitions constituents (MC).The contaminations of soil, surface water, sediment, and groundwater were mainly caused by past military activities, such as training for use of chemical and conventional weapons; practicing for decontaminating of equipment, vehicles, airplanes, roads, and buildings used in transporting, storing, distributing, and disposing of military munitions, fuel, and other related materials. Chemical warfare materials (CWM), unexploded ordnance (UXO), munitions debris (MD), munitions constituents (MC), and abandoned fuel facilities are safety and environmental hazards that may constitute an imminent and/or substantial danger to the public health, the environment, and ecosystems.

GIS technology has been used to locate impacted areas by analyzing historical aerials/drawing and LiDAR data, delineate extents of pollution on surface and underground through 2D and 3D modeling, assess environmental and ecological risks through data mining and analysis, help scientists and stakeholders to understand problems accurately through data visualization, design remedial strategies, and inform and educate the public. The massive GIS datasets collected and produced from these projects are securely stored, managed, and easily shared by various government agencies, industries, and the public through GIS web applications, with appropriate security and access controls. Furthermore, These GIS datasets can be re-used for future studies and investigations on these project sites instead of wasting money and time to re-collect, process, and analyze them.

2

Collecting Field Data More Efficiently with GIS and GPS Technologies

For many environmental projects, a large amount of field data are needed to locate and delineate impacted areas; characterize site conditions; evaluate current and potential future risks to the environment, ecosystems, and human health; design investigation and remediation approaches; and monitor, assess, and improve remedial processes and so on. Examples of commonly collected field datasets include samples; visual or equipment-aided site reconnaissance paths and observation points; roads and trails; topography and slopes; geological, geophysical, biological, and cultural surveys; soil types; historical, current, and future land uses; census data; properties and right of entry (ROE); wetlands; surface water bodies and streams; water wells; groundwater elevations and flow patterns; intrusive investigations; interviewing with local residents; and taking pictures.

Mobile GIS (e.g., ESRI ArcPad) and GPS (e.g., Trimble, Garmin) technologies can significantly increase field data collection efficiency, accuracy and security, and reduce or prevent human data entry errors. Mobile GIS and GPS can be programmed and customized to automate field data collection tasks by using project-specific data dictionaries, data capture and entry functions/tools, data QA/QC (quality assurance/quality control) and security procedures, in addition to synchronizing field data with office-based databases automatically through wireless communications. Trimble's TerraSync, Garmin's handheld GPS navigators, and ESRI's ArcPad were extensively used in the field data collection and mapping tasks in the case studies/projects which are discussed throughout this book. TerraSync, Garmin GPS, ArcPad and other related software/hardware are briefly reviewed below. Please refer to their manuals for detailed information.

Trimble GPS Pathfinder Office and TerraSync Software Packages

Trimble's GPS Pathfinder Office is an office-based software package of Global Navigation Satellite System (GNSS) data post-processing tools. It also supports H-Star data processing. TerraSync Studio utility within Pathfinder

Office can be used to develop and test customized TerraSync user interfaces to increase field data collection efficiency. The major features and tools of GPS Pathfinder Office software are briefly discussed here.

- *Data Dictionary Editor*: Data Dictionary Editor allows users to design a project-specific data dictionary or import one from a GIS based on its exact data schema and upload the data dictionary into mobile GPS data collecting devices/units. A data dictionary improves field data collection efficiency, accuracy and consistency. It ensures that data collected in the field will integrate seamlessly with existing GIS data in office. In addition to Data Dictionary Editor, GPS Pathfinder Office software also supports the development of conditional attribute data capture forms that dynamically adapt to previously entered attribute values for maximum data collection efficiency. In the field, the data capture form prompts users to select specific information from the predefined data dictionary instead of manually entering the information. This prevents possible data entry errors and reduces data entry time.

- *Post-differential correction*: The quality of field-collected GNSS data can be further enhanced by the post-processing differential correction tool of GPS Pathfinder Office software. Incorporating the DeltaPhase differential correction technology, the accuracy of data collected in the field can be significantly improved, reaching centimeter (1 cm or 0.394 in.) level, depending on the environment and the GNSS receiver.

- *Quality control of GNSS data*: Before exporting field-collected GNSS data to GIS, CADD, or other formats, data quality control features inside GPS Pathfinder Office allow users to analyze the GNSS datasets to ensure that they are complete and free of errors. For example, confirmed accurate background imagery and/or GIS data layers can be imported into GPS Pathfinder Office to verify the accuracy of the field-collected and post-processed GNSS data and detect possible errors. Users are also able to edit spatial data and their associated attributes as well as filter out unnecessary or unwanted GNSS positions to ensure high data quality before exporting.

- *Supporting a variety of data importing and exporting formats*: Data from a number of GIS and database formats can be imported into Trimble GPS Pathfinder Office, allowing existing GIS datasets to be taken back to the field for verification, updating, or adding new features. Field-collected and post-processed GNSS data/features can be exported into GIS, CAD, or other data formats with GPS Pathfinder Office.

Trimble's TerraSync is a data collection and data maintenance software package designed to work seamlessly with Trimble GPS receivers and Trimble

Positions Desktop Add-in for ESRI's ArcGIS software products. It is a mobile GPS software package. There are three TerraSync editions available, namely Standard, Professional, and Centimeter. The relatively low price, entry-level Standard edition is suitable for data collection tasks that do not require data update or maintenance. The Professional edition contains a variety of functionalities and tools required for data collection, update, and maintenance. It is ideal for field data collection tasks requiring decimeter accuracy and data update or maintenance. With an intuitive interface and user-defined pick lists, it can collect both spatial features (e.g., points, lines, and polygons/areas) and their associated attributes information accurately and efficiently. Existing GIS data layers can be verified, and updated in the field. It allows users to filter and sort datasets to identify the feature to be revisited. The Centimeter edition has all the functionalities and tools of the Professional edition, plus the ability to connect with a wide range of centimeter-grade Real Time Kinematic (RTK) GPS receivers to collect survey grade accuracy data.

RTK is a differential Global Navigation Satellite System technique, which is used to provide high-performance positioning in the vicinity of a base station, also known as a reference station. Using the signal's carrier wave phase measurements, instead of the information content (also known as code) of the signal, RTK utilizes reference base station(s) to provide real-time corrections for attaining up to centimeter-level accuracy.

TerraSync software integrates with a variety of GNSS receivers to meet different data accuracy (from meters to centimeter) needs and budget constraints. Not all field mapping and data collection projects demand very high positioning accuracy, which requires expensive hardware and software. TerraSync is also capable of using real-time differential GNSS corrections to improve data quality and accuracy in the field. TerraSync can operate on a variety of field mobile computers or handhelds, from the relatively affordable Trimble Juno series to the more expensive Trimble GeoExplorer series, the rugged Ranger, the Recon series, and the Nomad series, and so on. It can also integrate photo capturing into the data collection workflow using a handheld with a built-in digital camera. Photos are automatically attached to the current recorded feature and stamped with the time, date, and location the photo is taken.

Project-specific data dictionaries created in Trimble GPS Pathfinder Office software can be uploaded into TerraSync together with their associated spatial data layers for use in the field data collection. Data dictionaries improve data collection efficiency and consistency and reduce data entry errors through features such as attribute pick lists and conditional attributes. They help to preserve data integrity and comply with the GIS data structure. With the TerraSync Studio utility inside GPS Pathfinder Office software, TerraSync's user interface can be customized to enhance field data collection productivity and eliminate potential configuration errors, while minimizing the need for new staff training. The TerraSync Studio utility software provides a variety of functionalities and tools to develop and test customized TerraSync user interfaces to meet project-specific needs.

Accuracy-based logging settings specify the GNSS data quality that the enterprise GIS requires and the TerraSync automatically complies with it. To improve field data collection productivity users can utilize the Plan section to view a graphical prediction of the GNSS satellite constellation and identify the best time windows for data collection tasks, and schedule other activities in the time intervals not suitable for data collection.

Here are some key features of Trimble's TerraSync software.

- Providing an Accuracy-Based Logging capability to ensure that only data within the GIS accuracy requirements are logged.
- Delivering high data accuracy by using real-time H-Star technology, differential correction, and RTK GNSS receivers.
- Containing smart time-saving features, such as predefined pick lists, waypoint navigation, map-centric operation, integrated camera, and graphical status display.
- Transforming real-time corrected positions from the correction source datum to WGS84 for standardized logging of positions into the SSF file.
- Working directly with ESRI Shapefiles.
- Compatible with GPS Pathfinder Office software and the Trimble GPS Analyst extension for ESRI ArcGIS software for efficient data processing and differential correction.
- Supports Microsoft Windows Mobile and other Windows 32- and 64-bit operating systems.

Garmin Handheld GPS Navigators and GPS Utility Software

For field data collection tasks that do not require high data accuracy, Garmin's handheld GPS navigators can be an easy and affordable solution. Examples of such field data collection tasks include recording field observation locations, composite samples, reconnaissance routes/tracks, trails, streams, faults/fractures, wetland boundaries, geologic units, and so on. Garmin's Rino series handheld GPS navigators (e.g., Rino 530, 650, 650t) have reasonably good GPS capabilities and features. With BirdsEye Satellite Imagery and 100K topo maps preloaded as background imagery, a Rino series handheld GPS navigator is easy to navigate in the field. It can record up to 2,000 waypoints/locations, 200 routes/lines and 10,000 track points.

With GPS Utility software (GPSU), GIS data layers can be converted and uploaded into Garmin GPS devices. Vice versa, data collected with a Garmin

GPS can be downloaded and processed with GPS Utility. Freeware version GPS Utility can process up to 100 waypoints, 5 routes of up to 10 waypoints each, and 500 track points. A registered version GPS Utility can handle up to 65,000 waypoints, 500 routes, and 500,000 track points, which is adequate for most projects.

GPSU's key features are

- Exporting waypoints, route and track data from handheld GPS to PC in many file formats, such as GIS shapefile, Google Earth KML/KMZ, txt, DXF, and others
- Importing data in a variety file formats
- Basic editing capability for waypoints, routes and track points
- Entering coordinates in a variety of formats
- Using distance and bearing for waypoint creation
- Sorting waypoints by name/date/time/latitude/longitude/symbol/comments/altitude
- Filtering waypoints and tracks by name/geographic area/distance from current point
- Converting coordinates between map datums
- Viewing data in different coordinate formats (Lat/Long or Grids)
- Working with multiple sets of data at the same time
- Copying and pasting data between data sets even when referenced to different datums

GPS Utility supports many commonly used coordinate systems, such as the Geographic Coordinate System (GCS, also known as Latitude and Longitude), Universal Transverse Mercator (UTM), Universal Polar Stereographic (UPS), the Military Grid Reference System (MGRS, which is derived from UTM and used by NATO), USA State Planes (SPCS), and other continental or national coordinate systems. It supports a wide range of datums and grids too, such as those in Britain, Ireland, Germany, Switzerland, Sweden, Finland, the Netherlands, Belgium, Gauss-Boaga, the Czech Republic, Greece, and Costa Rica. It can also support datums and grids in West Malaysia, Singapore, Norway, Egypt, Georgia, Israel, New Zealand, St Lucia, and South Africa.

ESRI ArcPad Software

With integrated GIS and GPS capabilities, ESRI's ArcPad is a mobile field mapping, data collecting, and editing GIS software application. It can connect to personal or multiuser enterprise-level geodatabases, and check GIS data in and out. ArcPad can work directly with ArcGIS for Desktop, ArcGIS for Server and ArcGIS Online without using any Middleware to convert data back and forth. ArcPad supports the same vector and raster data formats, as other ArcGIS family software products do. Collecting data with ArcPad, GPS coordinates are automatically stored in GIS file formats, which can be seamlessly opened and used directly by other GIS software. When a spatial feature (e.g., a point, line, or polygon) is recorded, its data entry form is automatically displayed to collect the attribute's information. Drop-down menus, lists, check boxes, radio buttons and other features automate data entry, reduce or eliminate data input errors, and prevent missing data. Pictures can be taken in the field and linked to the recorded feature, all in one place. The use of electronic forms in the field significantly increases data entry efficiency, improves data accuracy through verification processes, and eliminates the tedious and error-prone work of re-entering paper-based information into a database in office. Data collected in the field can easily be uploaded into a master geodatabase in the office through the Internet via wireless communication. All other GIS mapping and internet applications connecting to the geodatabase are automatically updated with the uploaded data so that all stakeholders see the most recent data and make real-time informed decisions.

Similar to a desktop GIS mapping application, ArcPad allows users to visualize data layers easily by zooming to a specified layer, a spatial bookmark, the center of the current GPS position, or the extent of all visible layers. With ArcPad, users can also query spatial features and display their associated attributes; create a hyperlink to external files (e.g., photographs, documents, video, or sound recordings); measure distance, radius, and area; and calculate geographic statistics of selected features such as area, perimeter, length, coordinates, and total quantity. For better displaying and visualizing effects, the user can symbolize and label data layers. ArcPad also supports data editing and updating in the field through edit tools, such as creating, deleting and moving point, line, and polygon features in shapefiles. In addition, ArcPad enable users to add, delete, and move vertices for lines and polygons and append vertices to existing features. The coordinates of features are also editable using current GPS coordinates to replace less precise measurements.

ArcPad supports a wide range of commonly used projections/coordinate systems, such as Geodetic or Geographic Coordinate System, Albers Equal Area Conic, Cylindrical Equal Area, Double Stereographic, Transverse Mercator, Lambert Conformal Conic, Stereographic, and others. ArcPad is also capable of performing on-the-fly datum conversion from the geographic GPS input datum to the projection and datum of the current map. With many supported world datums in ArcPad, datum matching between GIS maps and GPS data becomes seamless.

ArcPad Studio is a desktop development framework for customizing or creating mobile GIS applications and tools. It is now included with the newer version of ArcPad 10.x. In the past, it was contained within the ArcPad Application Builder, a separate desktop software application. ArcPad Studio can be used to design custom data entry forms to streamline data collection in the field; create new toolbars that contain built-in and custom tools; build applets and scripts to solve specific data collection problems; develop extensions to support new file formats and positioning services; customize default ArcPad configuration files, and integrate digital cameras, monitoring devices, and other hardware/software into the data collection process. It can also be used to modify mobile GIS applications to suit special functionality needs, different user skill levels, and unique data collection requirements.

Data entry forms can contain multiple pages, any number of required and/ or read-only fields, and horizontal and vertical scroll bars. Tools can also be created to automate work flow in the field and aid the completion of forms. For example, a tool can be built to automatically fill in certain data fields, such as the project site name, field team leader, data collector, feature IDs, default values, coordinates (easting/northing, or latitude/longitude), data/time, and so on. It saves data entry time by setting the most frequently entered values as default values and have them automatically generated. Most of the time, the user does not need to do anything with default values. However, the default values can be manually changed when it is needed.

For example, in a sampling event, if it is known that most of the designed samples are surface soil samples, with only a few subsurface soils samples, surface water, and sediment samples, then the "Sample Medium" field's default value should be "Soil," and the "Sample Depth" field's default value should be "0 (feet/meter)." When a subsurface soil sample is taken, the "Sample Depth" field's value can be manually changed from default "0 (feet/meter)" to the actual sampling depth, while the Sample Medium" field remains as its default value of "Soil." When a surface water sample is taken, its "Sample Medium" field value can be manually changed from its default value of "Soil" to "Surface water." Drop-down lists or menus, check boxes, and radio buttons are used to allow users to select predefined data values instead of manually entering them to make data collection more efficient and prevent data entry errors. For a long-term sampling and monitoring project with multiple sampling events each year and hundreds or thousands of samples to be taken in each sampling event, these innovative data collection features can save a huge amount of time in entering data in the field, and checking and correcting data errors in office. They also simplify the work of field data collectors, making them happier, especially when working in bad weather and/or field conditions. This increases their morale and productivity.

Rules and triggers can also be built into forms and applications to ensure data quality, integrity and security. For example, certain fields can be defined as required fields, which enforce mandatory data entries. It is impossible to proceed to the next step or location until all the required fields are populated with correct data types and values. This prevents the possibility of missing data in the field and precludes the need to return to re-collect them, something which would negatively impact the project budget and schedule.

For another example, at a survey location, if its type is selected as a sampling location, then a sampling sub-form is displayed automatically by a built-in trigger to record detailed information of the sample location, such as sample ID, name of the sampler, date and time, medium (e.g., air, soil, surface water, sediment, or groundwater), depth/height, temperature, humidity, slope, vegetation and related other information. All required data fields have to be filled completely and correctly before it is possible to move to the next location.

Based on the requirements of a specific data collection project and/or the GIS skills of the user, ArcPad's default configuration file can be customized to adjust the ArcPad interface to better meet those requirements, such as hiding or eliminating unnecessary buttons and tools from the interface and adding new ones as needed. This helps to increase data collection productivity by reducing confusion and adding new functionalities and tools.

The customization files, such as applets, extensions, default configurations, and layer definition files, can be deployed to ArcPad by simply copying the customization files to the ArcPad's root directory folders on the end user's mobile device. When ArcPad is launched, it will automatically check the customization files in its root folder(s) and perform necessary actions accordingly.

The ArcPad Data Manager extension, included with ArcGIS for Desktop, can be used to check out feature classes, including relationships from a personal/file or multiuser enterprise geodatabase for use in ArcPad. Using these same tools in the ArcPad Data Manager extension, updated data feature classes can be checked back in to the geodatabase.

ArcPad supports a wide variety of handheld mobile GPS devices, such as the Trimble Juno series, GeoExplorer series, the rugged Ranger series, the Recon series, and the Nomad series GPS handhelds running in the Microsoft Windows Mobile operating system environment. With Trimble H-Star GPS technology, GeoExplorer GeoXH handheld can achieve real-time sub-foot (30 cm) accuracy with its internal antenna, and decimeter (10 cm) accuracy with a Zephyr or similar high-performance external antenna. GeoXH handheld offers Bluetooth and wireless LAN connectivity options. With its internal antenna, Trimble GeoXT handheld can deliver real-time submeter accuracy, which can be enhanced significantly by Trimble's GPScorrect extension software designed for ArcPad. The real-time data accuracy of the more affordable Trimble GeoMX handheld is in the 1–3 m range, with its internal antenna. Similarly, its data accuracy can be improved by the real-time differential correction of the GPScorrect extension. With advanced technology for extremely low multipath and low-elevation satellite tracking, the Trimble Zephyr external GPS antenna can achieve submillimeter accuracy.

The Trimble's GPScorrect extension for ESRI's ArcPad adds real-time differential correction capability to ArcPad for enhanced data accuracy in the field. With post-processing differential correction, accuracy of GPS positions can be improved from meter(s) to submeter, subfoot (30 cm) or better, depending on the data collecting environment and GPS receivers used, such as the highly accurate GPS Pathfinder ProXH receiver or the GeoXH handheld. While capturing data using ESRI's ArcPad software in the field, the GPScorrect extension automatically logs GPS positions and metadata (e.g., time) and differentially post-processes data in real time using the logged information.

Differential correction is a technique used to improve the accuracy of GPS positioning by comparing measurements taken by two or more receivers; that is, one or multiple stationary receivers (also referred to as base stations or reference stations), and one mobile receiver (also referred to as "rover"), which moves around to collect data. The stationary receiver or base station continuously records its fixed position over a control point. The difference between the stationary receiver's actual position and its calculated location is an estimate of the positioning error of that receiver at each given moment. A correction value is calculated to eliminate the error at that moment from GPS signals. The calculated correction from the selected base station(s) is later applied to the position recorded by the mobile receiver at the same moment. Using the logged information of the mobile receiver, all the recorded position data are processed using differential correction to enhance their accuracy. Among the selected base stations, GPScorrect software automatically evaluates and determines which source provides the best coverage, and connects with the best base station(s) to access the correction data.

With the Skyplot, Satellite Info, Plan, and other predictions of the satellite constellation functionalities and tools, Trimble's GPScorrect extension can also help users to check current GPS status, identify the best times of day for data collection, and plan other tasks in the time intervals not suitable for high quality data collection.

Trimble's GPS Analyst is an extension created for ESRI's ArcGIS Desktop software. Similar to the GPScorrect extension for ESRI ArcPad, GPS Analyst adds differential correction functionality to ArcGIS Desktop to further improve GNSS data accuracy in the office. It also allows users to specify the GNSS accuracy required for each feature class (e.g., point, line, and polygon), stores information about the quality of each GNSS position in the geodatabase, provides tools for analyzing this information and helps to fix or flag any exceptions/errors. With these features, Trimble's GPS Analyst extension helps to ensure field data quality. The Trimble GPS Analyst extension can also interact directly with personal or multiuser enterprise geodatabases, checking GIS data in and out. It can also work directly with data from Trimble's TerraSync software, which can be used as an alternative data collection method, as discussed earlier.

The Trimble's GPS Analyst extension can also import and export Trimble Standard Storage Format (SSF) files, which are used by Trimble data collection software products. Without GPS Analyst extension, Trimble's Pathfinder

Office software is needed to open SSF files and post-process the data. GPS Analyst extension can import SSF files into feature datasets in a personal or enterprise geodatabase. During the import process, SSF features are added to feature classes in the geodatabase, with GPS positions stored in new GPS sessions. When field data are imported into the geodatabase, GPS Analyst extension can differentially correct the GPS data and validate the estimated accuracy of the features. GIS data from a geodatabase or shapefile can be exported by GPS Analyst extension to an SSF file for use by other Trimble data collection software products, such as TerraSync GPS software. GPS Analyst extensions can be enhanced, customized, or extended with ArcObjects to better meet special data processing or other needs.

Once a field data collection mission is completed for the day, field datasets are downloaded to a laptop or desktop and undergo post-processing for extra precision using either Trimble GPS Analyst extension for ESRI ArcGIS Desktop software or the GPS Pathfinder Office software. After post-processing, editing and QA/QC checking, field-collected datasets are loaded back into (also known as "check in") the centralized master personal or enterprise geodatabase. All GIS applications (desktop, mobile and ArcGIS Online) connected to the geodatabase are automatically updated with the latest version datasets.

Case Study 2.1: Field Data Collecting, Processing and Mapping for Former Camp Sibert

As shown in Figure 2.1, the area formerly known as Camp Sibert is located in the Canoe Creek Valley between Chandler Mountain and Red Mountain to the northwest, and Dunaway Mountain and Canoe Creek Mountain to the southeast, in the foothills of the southern Appalachian Mountains. It mainly contains sparsely inhabited farmlands and woodlands, with an area of around 37,035 acres in a tract approximately 14 miles long and 5.5 miles wide. Interstate 59 (I-59) passes through the former Camp Sibert, and U.S. Highways 11 and 411 run parallel to the northeast and southeast, respectively. Several secondary roads connect to various parts of it. The general topography within the former Camp Sibert area varies from hilly uplands to flat lowlands. Elevation relief ranges from approximately 500 feet mean sea level (MSL) in valley to over 1000 feet MSL in the mountains along the site's boundary.

Geologically, the former Camp Sibert is in the southern portion of the Valley and Ridge Physiographic Province, which is a northeast-trending narrow belt of faulted and folded predominantly calcareous Paleozoic rocks. It extends for 1200 miles from the St. Lawrence Valley to the Gulf Coastal Plain in Alabama. The Valley and Ridge is characterized by valleys that are

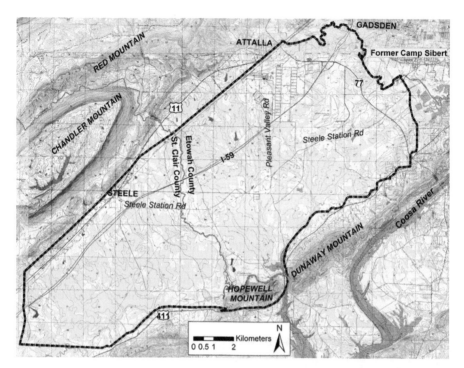

FIGURE 2.1
A topographic map showing general site setting of the former Camp Sibert.

underlain by limestones and dolomites with ridges that are more resistant to weathering and are capped by sandstones and cherty limestones and shales. The bedrock geology consists of the Cambrian Conasauga Formation, which is made of medium-blue-gray and fine-grained argillaceous limestone interbedded with dark gray shales.

A soil survey shows that the former Camp Sibert is within the Limestone Valleys and Uplands soil area of Alabama. Soils in this area were formed mainly in residuum weathered from limestone. Soils of the Tennessee River and Coosa River valleys were weathered from pure limestone and are mainly red clayey soils with silt loam surface textures. Most of the soils of the uplands are derived from cherty limestone. Bodine and Fullerton soils are extensive in many of these landscapes. They typically have a gravelly loam and gravelly clay subsoil, and a gravelly and silt loam surface layer.

The majority of the land of the former Camp Sibert is undeveloped and sparsely populated. Since 1949, most of the properties have been privately owned, either for farming, grazing, or as woodlands. The former airfield is now Northeast Alabama Regional Airport. The City of Gadsden has expanded into the area of the airport. There are some current and planned residential developments and improvements over the entire area.

Historically, the former Camp Sibert was established as the first large-scale chemical agent training area in the United States, used in preparation for the Second World War. In 1942, the Army acquired 37,035 acres of land there to develop a Replacement Training Center for the Army Chemical Warfare Service. Units and individual replacements received basic military instruction and training in the use of chemical weapons, decontamination procedures, and smoke operations, as illustrated by the historical photos taken from early 1942 to late 1945 in Figure 2.2.

Several types and calibers of weapons were fired at Camp Sibert, with the 4.2 in. mortar being the most-used heavy weapon. It was used for training in the use of white phosphorous and high explosive (HE) rounds. The most commonly used chemical agents at the former Camp Sibert are phosgene (a choking agent), mustard/nitrogen mustard (a blistering agent), and lewisite (a blistering agent). Some industrial chemicals, such as phosgene (CG), fuming sulfuric acid (FS), tearing agent (CNB), and adamsite (DM) might also have been used. In addition to chemical training, several types of conventional weapons were fired there too. Ammunition fired on ranges inside the camp included .30 caliber rifle bullets, machine gun rounds, .22 caliber rifle

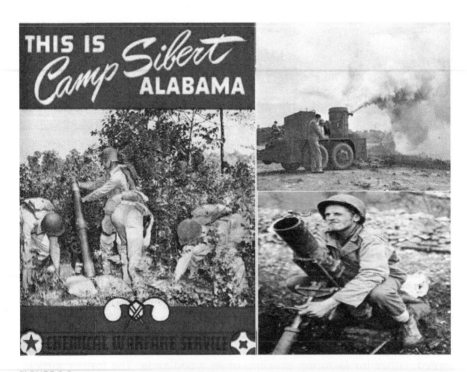

FIGURE 2.2
Historical photos of training activities in former Camp Sibert from 1942 to 1945. The 4.2 in. mortar shown in the lower right photo was used for both chemical and conventional munitions. It was the most commonly used weapon in the site.

bullets, sub-machine gun rounds, .45 caliber pistol rounds, grenades (both rifle and hand grenades), artillery, bazookas, and antiaircraft ordnance.

The chemical agent decontamination training at the former Camp Sibert included decontamination training on an airplane fuselage, walls, floors, different road types (gravel, concrete, sand, and macadam), shell holes, and trucks.

At the end of the Second World War those training activities were not needed anymore, and the camp was closed in late 1945. Some unwanted chemical warfare materiel (CWM) and training equipment were buried there. CWM and other unexploded ordnance (UXO), discarded munitions and explosives of concern (MEC), munitions constituents (MC), and munitions debris (MD) are safety hazards that may constitute an imminent and substantial danger to the public, project personnel, the ecosystem and the environment.

Due to the past chemical and conventional military training and burial activities, many areas (also known as sites) were contaminated by chemical materials, conventional weapon munitions, or both. There are over 35 identified CWM, conventional munitions and explosives sites inside Camp Sibert, as shown in Figure 2.3.

Since 1997, numerous environmental investigation and remediation projects have been conducted in Camp Sibert, such as engineering evaluation/cost analyses (EE/CA), remedial investigation/feasibility studies (RI/FS), and removal actions (RA). Hundreds of thousands of metallic anomalies have been identified and investigated, and many tons of munitions debris were retrieved from the contaminated soil and disposed in accordance with local, state, and federal regulations. More than 30 intact 4.2 in. mortars were also recovered. All intact munitions recovered were safely packaged, transported, and processed for neutralization. A large number of soil, sediment, surface water and groundwater samples were collected and analyzed for chemical agents; agent breakdown products; and hazardous, toxic and radioactive waste (HTRW) constituents.

For each environmental investigation or remediation project, its field work could last a few months, and an extensive amount of data could be collected, such as field observations, samples, reconnaissance tracks, geophysical meandering paths/towed arrays, anomaly points and polygons, survey/intrusive grids, bush cleaning status, right of entry (ROE) status, groundwater elevations, and geological surveys. An efficient, reliable, and high-quality field data collection process is crucial to the success of the project in terms of meeting or exceeding the project objectives and ensuring that the project is executed and completed on schedule and within budget.

After carefully evaluating the needs of the projects, ArcPad and TerraSync were selected for field mapping and data collection tasks. For ArcPad, custom data entry forms were designed with ArcPad Application Builder to streamline data collection in the field, including sampling locations, anomaly locations, geophysical survey status and results, intrusive investigation

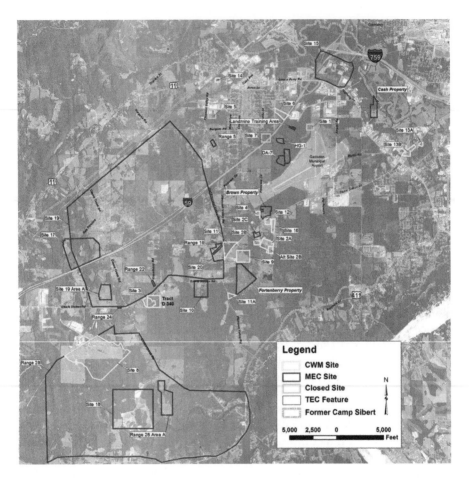

FIGURE 2.3

A GIS map showing CWM sites (the yellow polygons) and conventional munitions and explosives of concern (MEC) sites (the red polygons) within the former Camp Sibert. There are 35 CWM and MEC sites. Some sites were contaminated by both CWM and MEC.

grids and results, data QA/QC status, wetlands/flooded areas, ROE status, and bush clearance status. Tools, applets, scripts, data dictionaries, rules and triggers were built into forms and applications to improve field data collection efficiency and ensure data quality, integrity, consistency, and security, following relevant spatial data and metadata standards and guidelines. Trimble GeoExplorer series and other similar GPS handhelds were used in the field data collection. The GPScorrect extension was loaded with ArcPad for real-time differential correction to enhance GPS data accuracy in the field.

Field GIS data and geophysical survey results were synchronized to the centralized database through a secured wireless network. In the office, the GPS Analyst extension for ArcGIS Desktop was also utilized to further improve GNSS data accuracy. Processed, analyzed and updated GIS data—such as daily field work status; new sampling and geophysical survey plans; and updated ROE status—were synchronized/uploaded back into the field mobile GIS devices and geophysical equipment automatically. Each day, all the interested parties including the project team, field crews, GIS team, the client, regulators, and other related stakeholders, could see the same newly updated datasets and maps in real time, as illustrated by Figure 2.4.

From the centralized database, the project managers and their staff can query out detailed data tables and/or summary tables, create graphs, and generate reports. From the Geodatabase, the GIS team could extract, process, and analyze the geospatial data layers to generate GIS maps to be included in project reports and presentations. It can also be used to prepare final GIS deliverables to the clients.

In Figure 2.4, the GIS map in the lower right corner displays the geophysical investigation results at Site 8, a large CWM site inside the former Camp Sibert. The lines are the geophysical meandering paths. The color-coded dots are geophysical anomalies. The photos show the findings of the UXO (i.e., 4.2 in. chemical mortars), MD/MEC items and cultural debris on the site during the field investigation. The GIS map on the bottom is the Site 8 Removal Action daily project status map. It was automatically updated with the latest field investigation data each day. The green squares on the

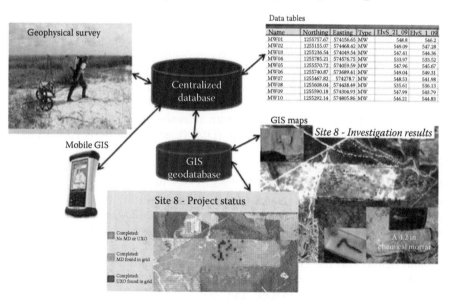

FIGURE 2.4

Field data was collected efficiently and synchronized to the centralized database through a secured wireless network and viewed by project team, field crews and other relevant stakeholders in real time. Detailed data and summary tables could be queried from the central database and maps and graphics generated from it. Updated data could be automatically synchronized/uploaded back to field mobile GIS devices and geophysical equipment. Procedures were built into the process to ensure data quality, integrity and security.

map represent completed grids with no MD or UXO, the golden squares being completed grids with MD found in them, and the red squares being completed grids with UXO found in them. The size of each grid is 100 ft. by 100 ft. About 3000 grids were investigated during the 24-month RA field investigation, from which a large amount of data was collected and many GIS maps generated.

Some of the GIS maps were published on the Internet so that people without GIS knowledge or software were able to view the maps and data, download them, and even edit them, depending on the access privilege levels granted to them. With these innovative and high-tech GIS approaches, all the interested parties could view the project data and maps at the same time and make the most informed, updated and appropriate decisions in real time.

Case Study 2.2: Data Collecting, Processing, Managing and Mapping for Three Large-Scale Regional and Nationwide MMRP Site Inspections (SI) Related Projects (Total: 1063 Sites)

In 2001, the United States Department of Defense (DoD) developed the Military Munitions Response Program (MMRP) to protect the public, the ecosystem and the environment from the potential hazards associated with the past military operations. Its remit especially included the use of military munitions in training and testing. The MMRP's primary responsibility is to address munitions-related concerns, including explosive safety; environmental and health hazards from releases of unexploded ordnance (UXO); discarded military munitions (DMM); and munitions constituents (MC) found at locations other than active operational ranges, such as Base Realignment and Closure (BRAC) installations and Formerly Used Defense Sites (FUDS) properties. The MMRP addresses nonoperational rangelands with suspected or known hazards from military munitions and explosives of concern (MEC) which occurred prior to September 2002 but which are not already included with an Installation Response Program (IRP) site cleanup activity. The MMRP program prioritizes sites for cleanup based on risk to human health and the environment. The U.S. Army Environmental Center (USAEC) is primarily responsible for completing the inventory for all active, BRAC, excess property, FUDS, and state-owned-and-operated National Guard Bureau installations, as required by the 2002 National Defense Authorization Act (NDAA). The US Army's Corps of Engineers is the executing manager for the inventorying project.

The three military munitions-related site inspections and scoping projects discussed in this chapter were conducted under the MMRP. There are over 1000 sites under inspection in these three large-scale nationwide MMRP projects. For each site, a sufficient amount of data was required to be collected. The data were used to assess the current and future risk of the site to human health, the ecosystem, and the environment, and to evaluate whether the site poses a significant threat to the extent that detailed investigation or immediate response is needed. Large quantities of high-quality data were crucial in conducting the risk analysis and making the decisions for each site. With so many project sites, so many people involved and limited project budgets and schedules, high quality, efficient and affordable data collection strategies and methods had to be designed and implemented to ensure data quality, quantity, consistency, integrity and security. It was a massive and challenging task.

Considering the special data needs and the budget/schedule constraints of these three MMRP site inspection projects, it was decided that ArcPad and TerraSync with high-performance GPS handhelds were the best solutions for collecting the datasets with a high accuracy requirement. These included locations of unexploded ordnance (UXO), sampling locations, parcels, and ROE. Custom data entry forms were designed to collect various datasets; improve data collection productivity; and ensure data quality and consistency, which are especially important to these large-scale projects. A variety of tools, applets, scripts, data dictionaries, rules and triggers were built into the forms and applications, following relevant spatial data and metadata standards and guidelines. The data collection hardware devices selected were the GeoExplorer series and other similar GPS handhelds. GPScorrect extension for ArcPad was used for real time differential correction to enhance GPS data accuracy in the field. GPS Analyst extension for ArcGIS Desktop was also used to further improve GNSS data accuracy in the office.

Due to the large number of project sites and budget/schedule constraints, the relatively cheaper Garmin's handheld GPS navigators were selected to collect datasets which required relatively less data accuracy, such as observation locations, reconnaissance transects, trails, habitats of protected species and cultural sites. Garmin's Rino series handheld GPS navigators (e.g., Rino 530, 650, 650t) were extensively used because they have reasonably good GPS capabilities and are easy to use for field staff with less GIS and GPS technical skills. They have BirdsEye Satellite Imagery and 100K topo maps preloaded as background imagery to help users navigate easily in the field. They can record up to 2,000 waypoints/locations, 10,000 track points and 200 routes/lines. GPSU software was used to process and upload/download Garmin GPS data.

For each project site, field-collected datasets were carefully processed, edited, QA/QC reviewed and then synchronized to the centralized master database. All stakeholders, such as the project teams, field crews, GIS teams, clients, regulators and other interested parties could view and use the same up-to-date datasets and maps in real time to perform data analysis and make decisions.

Project 1: FUDS MMRP Site Inspections (SI), SE and SW Regions (Total: 625 Sites)

The objectives of the FUDS MMRP SI are to determine whether the site poses a significant enough threat that a detailed investigation is needed; determine if an immediate response is needed; and collect data to prioritize the site for future cleanup action. From 2002 to 2013, more than 625 FUDS MMRP SI project sites located in southwestern and southeastern United States were investigated under this FUDS MMRP SI project. The project sites are located in the states of Alabama, Alaska, Arizona, Arkansas, California, Florida, Georgia, Hawaii, Kentucky, Louisiana, Mississippi, Nevada, New Mexico, North Carolina, Oklahoma, South Carolina, Tennessee, Texas, Utah and West Virginia, as well as the territories of Guam, the Northern Mariana Islands, American Samoa, Saipan, Puerto Rico and the Virgin Islands. Figure 2.5 shows the FUDS MMRP SI project sites in the southwestern region of the United States, and Figure 2.6 shows the FUDS MMRP SI project sites in the southeastern region of the United States and its territories.

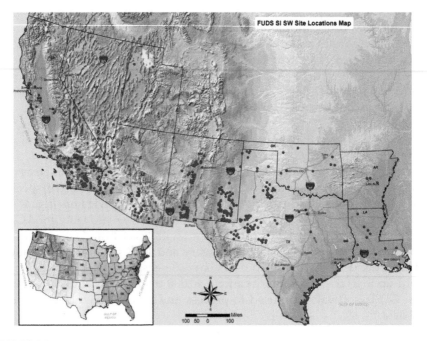

FIGURE 2.5

A GIS map showing the project sites (red points) of the FUDS MMRP Site Inspections (SI), in the southwest region (SW) region of the United States, including the states of Arizona (AZ), Arkansas (AR), California (CA), Louisiana (LA), Nevada (NV), New Mexico (NM), Oklahoma (OK), Texas (TX), and Utah (UT). These SW region states are highlighted in yellow in the inset map of the United States.

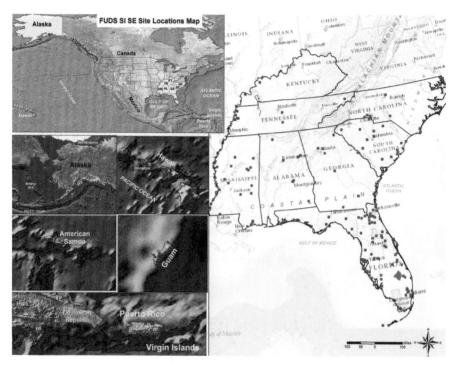

FIGURE 2.6

A GIS map showing the project sites (red points) of the FUDS MMRP Site Inspections (SI), in the southeast region (SE) region of the United States and its territories, including the states of Alabama (AL), Alaska (AK), Florida (FL), Georgia (GA), Hawaii (HI), Kentucky (KY), Mississippi (MS), North Carolina (NC), South Carolina (SC), and Tennessee (TN), and the territories of Guam, the Northern Mariana Islands, American Samoa, Saipan (north of Guam), Puerto Rico, and the Virgin Islands. These SE region states are highlighted in yellow in the inset map of the United States.

Project 2: U.S. Army National Guard Bureau NDNODS
Site Inspections (SI), Eastern Region (247 Sites)

The Non-Department of Defense Owned and Non-Operational Defense Sites (NDNODS) that were exclusively used by the Army National Guard and never owned, leased or otherwise possessed or used by the Army or other DoD component, are a subcategory of MMRP sites. In 2005, the U.S. Army National Guard (ARNG) was authorized to initiate an inventory of NDNODS to fulfill the Preliminary Assessments (PA) phase under the Comprehensive Environmental Response, Compensation and Liability Act of 1980 (CERCLA). ARNG completed the NDNODS inventory in 2009. The NDNODS were authorized as eligible for funding on the Defense Environmental Restoration Program (DERP) on 26 February 2010. Site Inspection (SI) is the next applicable phase in the CERCLA process for MMRP sites designated as eligible

NDNODS and other State Army National Guard MMRP sites. Each State or territory Army National Guard is considered a separate "installation" and may include multiple sites within the state or territory. There are 247 NDNODS SI sites in eastern region, under this project in 20 states and two territories, as shown in Figure 2.7.

The primary goal of the NDNODS SI is to collect the appropriate amount of information necessary to make the following decisions for a site: (1) whether or not a remedial investigation/feasibility study (RI/FS) is required; (2) whether or not an emergency or time-critical response action (TCRA) is required; and (3) whether the site qualifies for no further action (NFA). The installation-wide SI will address MEC hazards (including both UXO and DMM), as well as munitions constituents (MC) issues for the MMRP-eligible

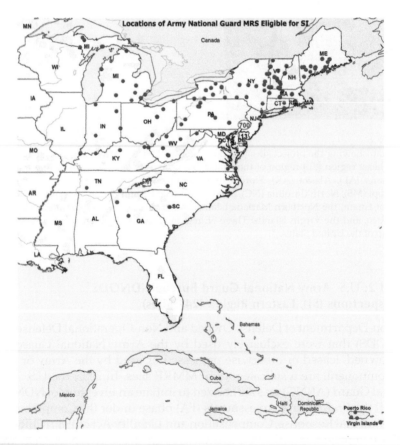

FIGURE 2.7
Site location map of the Army National Guard Bureau NDNODS Site Inspections (SI), in the Eastern Region of the United States. The red points represent the approximate locations of the ARNG NDNOD SI sites, also known as Munitions Response Sites (MRS).

sites. The secondary goals of the SI are to collect the necessary information to complete the Munitions Response Site Prioritization Protocol (MRSPP) for each MMR site and allow the development of better Cost to Complete (CTC) estimates for the MMRP-eligible sites by documenting remaining liabilities and the expected remediation strategy.

Project 3: Nationwide Chemical Warfare Materiel (CWM) Scoping and Security Study (Total: 91 Sites)

The objectives of the CWM Scoping and Security Study (also known as the CWM Study) were to prioritize the FUDS-eligible suspect CWM project sites for future funding and actions; to involve the public, federal, state, tribal, and local stakeholders in the decision-making process for determining potential further action; and to identify security and safety concerns. There were 91 CWM sites selected for this project, located in 28 states and 2 territories, as shown in Figure 2.8. The main purpose of this project was to provide the background information required for selecting response actions for the sites where Chemical Agent Identification Sets (CAIS) were the sole type of potential CWM. CAIS were training aids used for training personnel in the identification and decontamination of the chemical agents and industrial chemicals used in chemical warfare. Since historical records are incomplete for these sites, especially lacking information about the final disposal site of the CAIS, it is possible that buried CAIS may remain at the undeveloped sites. This project also collected background information about previous investigations at CAIS sites; explained the technological limitations of locating CAIS; evaluated the contaminant fate and transportation of the chemical agents and industrial chemicals in CAIS; performed risk evaluations of both human health and the environment; and evaluated potential remedial action alternatives for the sites.

The determination of a suspected CWM site is based on an evaluation of its history and by investigations conducted at the site. Data from studies such as the Inventory Project Report (INPR), the Archives Search Report (ASR) or the CWM Scoping and Security Study may be used to categorize a site. Some sites might have had CWM before but are considered to no longer have extant CWM based on historical documentation, removal or cleanup of CWM, or further evaluation.

The use of CWM by the United States started in 1917, during the Great War. The United States continued developing agents and training troops in the use of CWM through to the Second World War in the 1940s. In 1969, the United States renounced first use of chemical and biological weapons. Since then, research and development were limited to defensive agents and antidotes. Some FUDS were used for the development, production, and storage of chemical weapons as well as for the training of troops. The histories of the 91 sites selected for this study show a wide range of past conventional ordnance and CWM-related uses, activities and training. The sites fall into

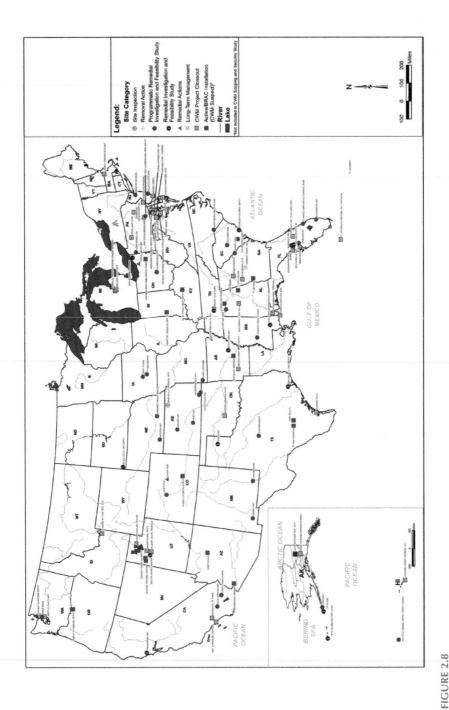

FIGURE 2.8

A GIS map showing locations of the 91 CWM sites selected for the scoping and security study in the United States and its territories.

four general categories: (1) Military Airfields; (2) Forts, Camps, and Ranges; (3) Production Plants; and (4) Arsenals and Depots.

A combination of these three MEC and CWM MMRP SI projects yields 1063 sites across the United States and its territories. For each site, a sufficient amount of data was required to be collected, stored, documented, processed, analyzed, and presented to meet the objectives of the site. GIS and GPS technologies were extensively used in the collecting, processing, editing, storing, managing and analyzing of the massive datasets; supporting field investigations; assisting public meetings; and the writing of the various reports by supplying figures, graphics, large posters, summary tables and data calculations. A final GIS deliverable, containing all GIS datasets and maps, was also required to be prepared and submitted for each site in compliance with the Spatial Data Standards for Facilities, Infrastructure and Environment (SDSFIE); the Federal Geographic Data Committee's (FGDC) Content Standard for Digital Geospatial Metadata (CSDGM); the Spatial Data Transfer Standard (SDTS); the Geospatial Positioning Accuracy Standards, the Geographic Information Framework Data Standard; and the USACE Data Item Description (DID) for Geospatial Information and Electronic Submittals (WERS-007.01).

Before starting the field investigation of a project site, GIS was used to gather as much background information as possible to prepare work and safety plans. This work would include obtaining right of entry (ROE) for the affected properties; locating possible UXO, MEC and MD items; locating habitats of protected species and wetlands; and designing site layouts, sampling locations, instrument-aided visual survey paths (also known as representative qualitative reconnaissance tracks, or transects) and possible brush clearance. Commonly needed background datasets include topography and slopes, as shown in Figure 2.9; recent and/or historical aerial photos, as shown in Figure 2.10; LiDAR imagery; roads and trails; population; properties and ROE; water wells, wetlands, habitats and cultural sites; historical, current and future land uses; and soil, geology hydrology. These background datasets and proposed samples, visual survey paths, locations of suspected MEC items and others are processed and uploaded into mobile GIS and/or GPS devices for field investigation uses. As discussed in the first case study, mobile GIS and GPS were programmed and customized to make field data collection more efficient, accurate and consistent.

For some special sites, public involvement was required at certain stages of the investigations. GIS was used to select the relevant properties, compile mailing lists and generate large posters, graphics and summary tables for displays, presentations or media releases.

Some of the collected datasets, such as UXO or other MEC items on site; water wells; census data; land use; wetlands; and surface and groundwater habitats, were used to evaluate risks on human health, the ecosystem and the environment. Below are two example maps (Figures 2.11 and 2.12)

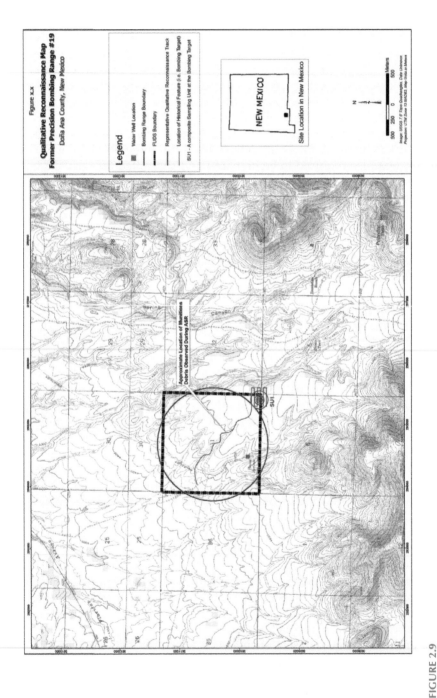

FIGURE 2.9

A work plan map with a topographic imagery background, showing the FUDS (the dashed black line polygon), a bombing range (the red circle), a water well (the blue square), a proposed sampling unit (SU), and the representative qualitative reconnaissance tracks (the purple lines).

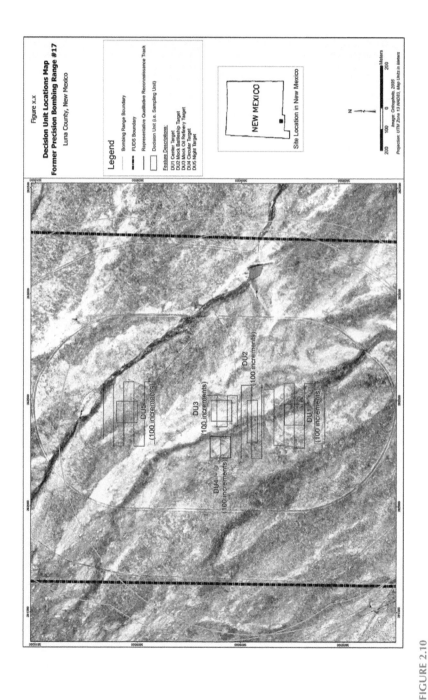

FIGURE 2.10

A work plan map with a high-resolution aerial photo background, which clearly shows the historical bombing ranges (the two large ellipses oriented north–south) and various bombing targets (listed in the legend) inside the ranges. Based on the information of these features, composite sampling units (red squares) and representative qualitative reconnaissance tracks (purple lines) were designed.

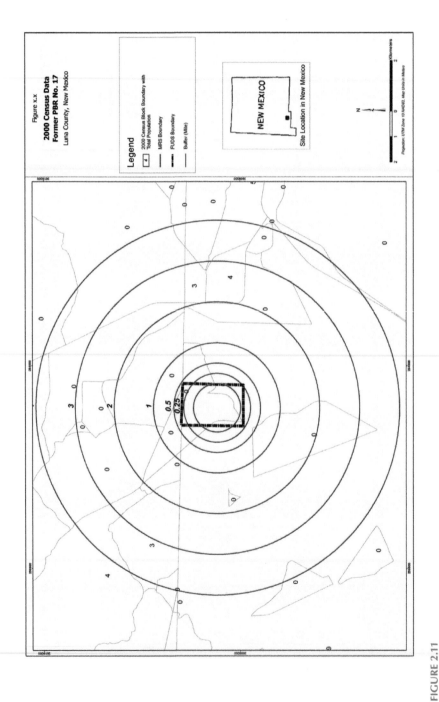

FIGURE 2.11

A GIS map showing 2000 census tracks and the population within 4 miles of a former precision bombing range (PBR) FUDS SI site. This map and data were used in the site risk assessment and in the SI report.

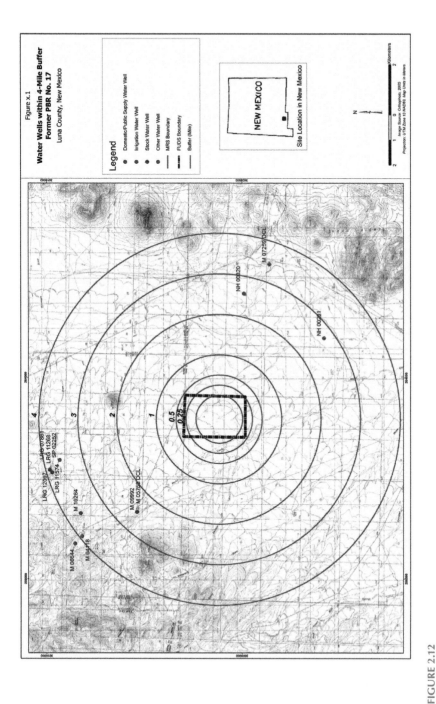

FIGURE 2.12

A GIS map containing groundwater wells within 4 miles of a former precision bombing range (PBR) FUDS SI site. This map and data were used in the site risk assessment and in the SI report.

prepared for the risk assessments. Figure 2.11 shows the census data of the project site and Figure 2.12 shows the water wells within 4 miles of the site. They were also included in the SI report of the investigated site as a matter of course.

During a site investigation, various datasets, such as samples; observation points and items discovered (UXO, MD, cultural debris, habitats, and others of interest); photos; and visual survey paths, were collected. This information synchronized automatically to centralized office-based databases, as illustrated in Figure 2.4. GIS and GPS tools were used to process and edit field GIS and GPS datasets which were then securely stored in GIS databases. Updated GIS datasets, such as new ROE, additional or revised sampling locations and reconnaissance tracks and brush cutting, were synchronized back to field mobile GIS and GPS devices to guide fieldwork activities. With the automated or optimized systems of data collecting, processing, editing, storing, mapping and sharing, project managers and their staff, field teams, clients and other relevant stakeholders could view and use the datasets and maps in real time and make educated decisions.

After a site investigation was completed, the final version of field datasets and other related data were processed, analyzed and compiled into figures to be included in the SI report. Figure 2.13 is an example report figure. Last, but certainly not least, a final GIS deliverable, consisting of all project datasets and maps, was prepared and submitted following all the relevant spatial and metadata standards and regulations, such as the SDSFIE, FGDC CSDGM, SDTS and the USACE DID for Geospatial Information and Electronic Submittals (WERS-007.01).

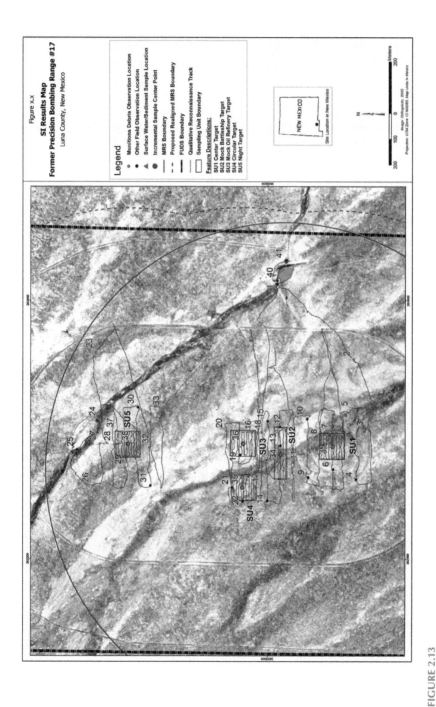

FIGURE 2.13

A final SI report figure showing the actual composite sampling units (black squares) and representative qualitative reconnaissance tracks (purple lines). They are slightly different from the proposed ones shown in Figure 2.10 due to field conditions, such as terrain, vegetation, barriers, and so forth.

Case Study 2.3: Locating and Destroying Final Two Caches of Chemical Weapons in Libya

This was a special, high-profile project carried out from 2013 to early 2014. On February 2, 2014, the *New York Times* reported on this project in great detail in an article called "Libya's Cache of Toxic Arms All Destroyed." Due to the sensitivity of this project, the background information here is based solely on the *New York Times* report. The location and other specific information about this project will not be discussed in this book either.

As described in the report, two additional caches containing around two tons of mustard agent were discovered by the new Libyan government in November 2011 and February 2012. They were not declared by Colonel Qaddafi's former government. Unlike other chemical warfare materiel (CWM) that had been declared—and which was mainly mustard agents stored in large containers—the newly discovered hidden caches were already armed and loaded into 517 artillery shells, 45 plastic sleeves for rocket launch, and eight 500lb bombs. They were combat-ready chemical weapons and the new Libyan government had no way of destroying them. Therefore, Libya requested technical help from American and European allies and the relevant United Nations agencies.

The newly discovered caches alarmed the international community due to the chaotic security situation in Libya at that time. There was an urgent desire to accurately locate and destroy the chemical weapons as quickly as possible to prevent them from falling into the wrong hands. The Pentagon and its Defense Threat Reduction Agency initiated a special project under the Nunn-Lugar Cooperative Threat Reduction program. In this, they contracted an engineering company to work with the Libyan government to oversee the building of the disposal site and to supervise the locating and safe destruction of all chemical weapons.

GIS image analysis tools were utilized to efficiently process a large amount of satellite and aerial images to precisely locate chemical weapons storage facilities (mainly bunkers) hidden in the vast desert. With limited information and a tight schedule, it was a very challenging task, like finding a needle in a haystack.

There are a number of differing theories and methods used to identify features through image interpretation. One of them is based on the property of pixel similarity. In a digital image, each pixel has a unique value representing its physical property. Adjacent pixels with similar values are grouped together and separated from their neighboring pixels with significant different values to form a pattern. Through analyzing similar and different patterns, features in an image are identified and recognized. For example, the values of the pixels of a paved road segment are similar among themselves, but they are dramatically different from the values of the pixels of the surrounding soil. These contrasting differences form the boundaries of the patterns/features.

Another common method of identifying and extracting features from image analysis is thresholding. In this approach, a range of pixel values is used as the threshold for a type of feature in order to separate it from the background image data. There are many technical approaches in this category, such as Clustering, where the gray-level samples are clustered in two parts as foreground (the feature) and background; the Histogram shape method, where the feature's histogram shapes are analyzed; and the Spatial method, which uses higher-order probability distribution and/or correlation between pixels.

Based on human knowledge of the features of interest, GIS pixel classification functionalities are used to select and categorize features by grouping similar pixels into the same feature class. Due to similarities between some features, it would be challenging to assign appropriate values to these feature classes. Knowledge, experience and experimentation are important to achieve satisfactory image analysis results.

If the interested features are needed in a vector data format, GIS raster-to-vector conversion tools can be used to convert the extracted image features. Depending on the quality of the images, the nature of the features of interest and the image processing, analysis, and categorizing processes, the converted vector features need to be cleaned and edited to certain degrees to correct the errors introduced during the image analysis and data conversion processes.

For this special project, the image analysis process was significantly automated. This meant it could work through the huge amount of images that covered such a large desert region to identify features such as roads, disturbed areas, vehicles, fences, gates and the hidden chemical weapons storage bunkers. Below are a few example maps showing the features of interest as identified from the massive image analysis.

Figure 2.14 shows the general location of the chemical weapons storage facilities (i.e., bunkers), a cache of unexploded ordnance (UXO), damaged vehicles, tanks and other items of interest. A section of the highway's name was removed due to the sensitivity of the site. Figure 2.15 is a work area and security layout map. Figure 2.16 is a close-up view of a few selected bunkers and their safety buffers (front, 12 m; side and back, 6 m). Figure 2.17 shows detailed information on the layout of chemical weapon disposal sites and pictures of the bunkers.

Using GIS, GPS and other relevant techniques and tools, all the chemical weapons and related items were located. They were completely destroyed using a custom-built static-detonation chamber supplied by a Swedish company. Operating like a large oven, the dangerous chemical materials inside the chamber were vaporized at extremely high temperatures between 750° and 1000°F. The gases created in the process were scrubbed by special

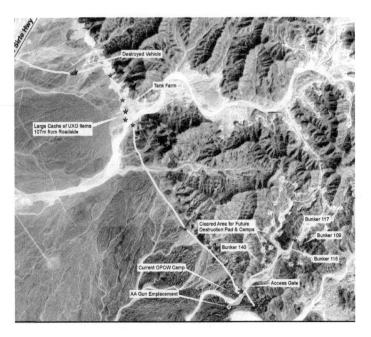

FIGURE 2.14
General location of the project area. The yellow lines are access roads. The bunkers were used as chemical weapon storage facilities. The red stars are UXO items.

FIGURE 2.15
Work area and security layout map. The thick green line is the security patrol path. The red line is the work area. The dark purple lines are access roads.

FIGURE 2.16
A close-up view of a few bunkers. The purple line is an access road. The red lines polygons are the safety buffers (front 12 m; side and back, 6 m) around the bunkers.

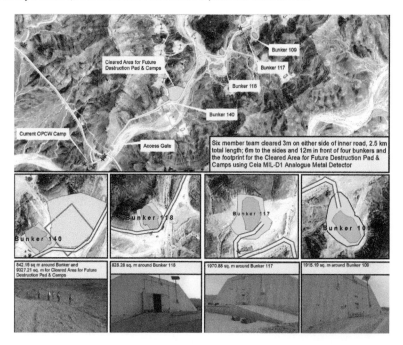

FIGURE 2.17
Pictures of the bunker, the cleared areas around the bunkers, and one large cleared area for the destruction pad and camps used in the disposal of the chemical weapons. The red stars are UXO items.

filters. Inspectors from the Organization for the Prohibition of Chemical Weapons supervised the destruction.

It was a successful global cooperative threat reduction effort led by the United States and supported by other countries and international agencies. According to the *New York Times* report, Canada contributed funding to restore water, sanitation services and electricity to the project site and to build living quarters for the Western and Libyan contractors. Germany lent assistance by transporting international inspectors to the site; the Libyan contractors were also trained in Germany and Sweden. It was highly praised by the Organization for the Prohibition of Chemical Weapons, Green Cross International, and the U.S. government.

The technologies tested in this successful project—in addition to the experience gained and the lessons learned—would undoubtedly be helpful in future efforts to eliminate chemical warfare materiel and/or weapons in other countries or regions. This would serve to make the world a safer place to live and thrive for the generations to come.

3

Processing and Analyzing Data to Extract Useful and Meaningful Information

Relevant and sufficient amounts of data are required to make correct and accurate decisions. Although some data are straightforward and easy to understand, other data—especially those with large quantity and sophisticated relationships to other data—could appear to be complicated, intimidating, and confusing. These data need to be processed and analyzed to reveal the patterns and relationships hidden in the large amounts of numbers. For example, while reviewing the historical records and reports of previous investigations at an environmentally impacted site, an environmental scientist gathers a large amount of datasets, such as historical drawings and aerial photos; chemical results of soil, surface water, sediment and groundwater samples; current and planned future land uses, hydrology, geology, soil, vegetation and elevation data; geophysical transects and anomalies and intrusive investigation results; water wells; wetlands; census data; cultural sites, habitats of protected species; affected properties; right of entry (ROE) information, and so on.

However, by looking at these datasets alone, it could be hard to locate the environmentally impacted areas, delineate their extents on the surface and underground, visualize their patterns and relationships, and understand their changes over time. It could also be difficult to evaluate their current and possible future risks to human health, the environment and the ecosystem and to design remedial investigation strategies and detailed technical approaches. Scientists and decision-makers could be confused by these large quantities of datasets piled up on their desks. No matter how valuable these datasets are and how much money and time are spent collecting them, if data users are not able to fully understand them and visualize their relationships, patterns, and trends, the values of the data are not fully utilized. They will not help much in precisely understanding problems and finding solutions. In extreme cases, if data users are confused or misled by the large and complex datasets, they might make inaccurate or even wrong decisions. Therefore, more data do not always result in better decision-making.

A Geographic Information System (GIS) offers powerful data processing, analyzing, and modeling tools to help scientists and decision-makers to visualize and understand data better, to get useful and more meaningful information from them, and to make sound decisions when solving problems. For example, with GIS, a historical aerial photo showing features of former air-to-ground practice bombing ranges, targets, and bombing craters can be georeferenced and overlain onto a current map. This enables locating the areas impacted by the bombing-related military munitions, including unexploded ordnance (UXO), munitions debris (MD), and munitions and explosives of concern (MEC). It also helps in delineating their boundaries and designing investigative and/or remedial approaches, such as obtaining ROE to the affected properties; planning brush clearance; deciding sampling types and locations; proposing geophysical survey methods, areas, and transects; designing intrusive investigation grids and/or trenches; and installing monitoring wells.

Similarly, a historical drawing showing the detailed layout of an installation can be scanned and georeferenced to identify potential polluted areas and their possible sources, such as ground and underground fuel or munitions storage facilities; fueling stations; hazardous materials loading, unloading and processing areas; maintenance shops; landfills/dump areas; and burial pits. Analytical results of samples can be statistically analyzed and displayed as color-shaded contour maps showing concentration patterns throughout the whole project site and changing trends over time. Underground soil and water contaminations can be delineated and illustrated using three-dimensional (3D) modeling and visualization techniques. Risks to human health, the environment, and the ecosystem can be evaluated by analyzing the relationships between the related data layers, such as samples, census data, water wells, groundwater aquifers, wetlands, rivers, streams, lakes, oceans, parcels, habitats, geology, soil, land covers and uses, and topography. By statistically analyzing existing data coverage of a project site, data clusters (i.e., too much and overlapping data areas) and data gaps (i.e., insufficient data coverage areas) can be identified and situations in adjacent areas can be predicted. For data gaps, additional data collection efforts are required to achieve sufficient data coverage of the whole project site to make accurate decisions. In areas with data clusters, monetary savings can be achieved by avoiding new data collecting efforts there and by reducing some existing data collection activities, for example, decommission unnecessary monitoring wells and remove unwanted sampling locations.

Case Study 3.1: Statistically Analyzing Existing Geophysical Survey Data to Predict Possible Anomalies in Adjacent Unknown Areas

Site 8, located in the south central portion of the former Camp Sibert, was a 375 acre area within the former Range R30 that had been impacted by toxic chemical munitions. Topographically, Site 8 is characterized by relatively flat lowlands and floodplains. The general vegetation of Site 8 consists of mixed pine and hardwood forest. The northern section of Site 8 is covered by a thick intermediate pine forest; its central section was cleared for grazing land, and its southern section is undeveloped and covered by thick pine and hardwood forest.

From early 1942 to late 1945, during the Second World War, Range R30 was constructed and used as a toxic gas 4.2 in. chemical mortar training range. Site 8 was the impact area within Range R30. Inside the 4.2 in. mortar range, a Japanese-style pillbox was constructed for jungle warfare and pillbox-attack training purposes. The 4.2 in. mortar was historically used in training to fire chemical or smoke rounds filled with mustard, phosgene, lewisite, tearing agents, white phosphorus, sulfur trioxide, chlorosulfonic acid solutions, etc. Training in Range R30 and Site 8 ended after the war in late 1945 and the site was closed. In June 1948, a chemical mortar round in the back of a pickup truck was accidentally detonated near Site 8 and the people inside the vehicle were injured with mustard-agent-type burns.

A chemical warfare materiel (CWM) removal action (RA) at Site 8 was initiated in June 2004 to remove and dispose of all recovered CWM, ordnance-related scraps, and explosives hazards from selected areas—totaling around 247.73 acres—of Site 8. The southern, heavily forested section of Site 8 was not included in that RA project.

Before this RA project was carried out, all the geophysical survey and intrusive investigations datasets from previous investigations were retrieved and reviewed. Project records show that during an engineering evaluation/cost analysis (EE/CA) investigation in 2000 the central portion of Site 8 was geophysically surveyed using meandering paths and 532 anomalies were identified. Another phased EE/CA investigation was conducted at Site 8 in 2002. The site was geophysically mapped using towed arrays, with 8673 more anomalies identified. The surveyed (mapped) geophysical anomalies of these two previous EE/CA investigations were combined (totaling 9205) and are shown as red dots in Figure 3.1.

GIS statistical tools and functionalities were used to analyze these surveyed geophysical anomalies to predict possible anomaly distribution patterns in the adjacent unknown and un-surveyed areas based on the Kriging geostatistical method.

The basic Kriging method was developed by the French mathematician Georges Matheron, based on the studies and experiments of Danie G. Krige, a South African mining engineer who tried to predict the most likely distribution of gold based on the samples from a few boreholes using some statistical calculation methods. Acknowledging Krige's pioneering work, this unique statistical analysis method was named after him. The Kriging method, also known as Gaussian process regression, is a widely used statistical data point interpolation technique that uses existing measured/known data values to calculate a predicted value in an unmeasured/unknown adjacent location based on the assumption that the spatial variation in the data being modeled is homogeneous across the entire study area.

FIGURE 3.1
Geophysical anomalies map of Site 8, former Camp Sibert. The blue line represents the boundary of Site 8. The cyan dashed line represents the boundary of the suspected impact area by the 4.2 in. chemical mortar firing training. The red dots are surveyed geophysical anomalies from previous EE/CA investigations of 2000 and 2002. The yellow dots are statistically calculated possible anomaly locations from the surveyed data.

There are other, similar regression-based statistical interpolation algorithms and methods, such as inverse distance squared, splines, radial basis functions, triangulation, and so on. They all predict the value at an unknown location as a weighted sum of data points at its surrounding locations. Weights are calculated and assigned based on the distance of the prediction locations from the measured points, directions, and the overall distribution patterns of the measured points of the area of interest. Distance is the most important factor. Weights decrease with increasing separation distance of the predicted point and its surrounding measured data points. What makes Kriging unique among interpolation methods is that it calculates and assigns weights based more on a data-driven weighting function, rather than an arbitrary function, and it also offers a way to estimate the variance of predictions, that is, estimation errors.

Regardless of the interpolation algorithm chosen, high-quality datasets usually yield more accurate prediction results, and vice versa. In general, if known data points are accurate, relatively dense, and uniformly distributed throughout the study area, the calculated predictions are more reliable. However, if the data points are scarce or in clusters with large data gaps in between, it will be hard to achieve an accurate prediction no matter which interpolation algorithm or method is used.

In Kriging, there are different methods for calculating the weights (i.e., Kriging variants), such as simple Kriging, ordinary Kriging, universal

Kriging, IRFk-Kriging, indicator Kriging, multiple-indicator Kriging, disjunctive Kriging, and lognormal Kriging.

Following are three main Kriging variants discussed in more detail.

- Simple Kriging: With a simple Kriging method, it is assumed that the mean is constant over the entire domain of interest. It is an unbiased weight estimating approach. It calculates the value of the prediction location based on a set of weights estimated from a set of neighboring data points. The weight on each data point decreases with increasing distance to the prediction point; that is, closer data points have more influence on the prediction point than the more distant ones. It also decreases the weights of data points in clusters, instead of treating them the same way as normally distributed data points. This is one of the advantages of Kriging when compared with other data point interpolation algorithms.
- Ordinary Kriging: Slightly different from simple Kriging, ordinary Kriging assumes that the mean is constant in the local neighborhood of a prediction point, instead of the entire study area.
- Universal Kriging: Quite different from simple Kriging and ordinary Kriging variants, in universal Kriging the mean can be nonconstant, allowing for local variations.

As Figure 3.1 demonstrates, the mapped geophysical anomalies, that is, the red data points, are fairly dense and uniformly distributed throughout the study area except for in the suspect impact area. This is the section inside the dashed cyan circle, where more anomalies were recorded. Even in this area, data points are relatively evenly distributed without any clusters or data gaps. Therefore, the simple Kriging method was selected to predict possible anomalies in the adjacent un-surveyed areas from the mapped geophysical anomalies. The results are shown as yellow points in Figure 3.1.

The distribution pattern/trend of the predicted anomalies corresponds very well with the pattern of the mapped anomalies, that is, there are more anomalies inside and near the suspect impact area and gradually fewer and fewer anomalies moving fartheraway from it.

Since the southern portion of Site 8 was not included in the scope of the RA project, the predicted anomaly points in that area were manually cleared from the map. This geophysical anomaly distribution map, showing both the mapped and predicted anomalies, was used by both the RA project team and the field crews to acquire right of entry (ROE), plan brush clearing areas, design additional geophysical survey areas, design intrusive investigation grids system, position vapor containment structures (VCS), and calculate safety exclusion zones. Using this GIS data analysis technique saved the project time and money.

Case Studies 3.2, 3.3, 3.4, 3.5: Analyzing Recent, Historical
Aerial Photos/Drawings and LiDAR to Identify and
Delineate Environmentally Impacted Areas

Case Study 3.2: Recent Aerial Photo Analysis

Encompassing approximately 4900 acres of land, Site 19 is the largest site
inside the former Camp Sibert. It was impacted by the conventional mortar
training conducted from 1942 through 1945. In 1995, during Phase II of the
Site Characterization project, a limited geophysical survey was conducted
and around 341 subsurface anomalies were identified. The Phase II Site
Characterization Report recommended additional geophysical mapping
along with intrusive investigations that would include excavating a a certain
number of the identified subsurface anomalies.

During a remedial investigation/feasibility study (RI/FS) project of Site
19 in 2002, the project team reviewed all previous investigations of Site
19, including the Site Characterization Report and the anomalies data.
Constrained by the tight project budget and schedule for this large site, innovative approaches had to be utilized to save the costs of the recommended
geophysical mapping and intrusive investigations. The most important
approach was to identify and delineate the areas of the greatest concerns
using GIS technology. The theory is that most ordnance and explosive (OE)
wastes, such as unexploded ordnance (UXO) and munitions debris (MD)
from conventional weapons, are usually on the surface or buried relatively
shallowly in the ground. Therefore, the areas disturbed by construction,
farming, or other activities are unlikely to still have OE wastes. So, investigations should be focused on the areas undisturbed since 1945, instead of
conducting geophysical mapping and intrusive investigations on the whole
large site. This was nearly impossible in any case with the limited project
budget and schedule.

To achieve the cost-saving goals and ensure the project ran on schedule,
GIS raster image analysis techniques were used to analyze the high resolution and relatively recent 1997 aerial photos to identify the undisturbed areas
inside Site 19. The 1997 aerial photos were selected for analysis because of
their high resolution and also because not much ground disturbance had
occurred in Site 19 since they were taken. Some 1950 aerial photos were also
analyzed to further enhance and confirm the 1997 aerial analysis results, as
shown in Figure 3.2.

As discussed in Chapter 2, there are numerous methods to identify and
extract features from satellite images, aerial photos, LiDAR data, scanned
images, and so on. One of the most commonly used methods is to classify
pixels based on their physical properties, that is, the values of the pixels.
Adjacent pixels with similar values are grouped or classified together by a
range of values and separated from their neighboring pixels with significantly

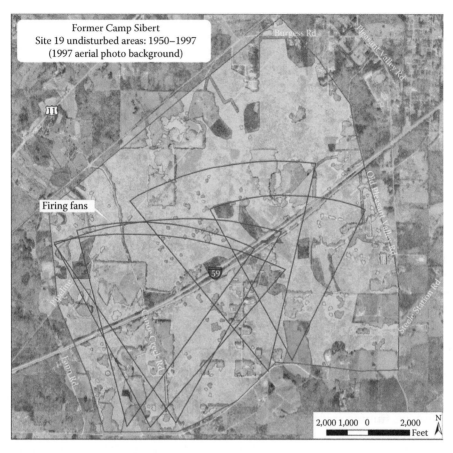

FIGURE 3.2
1997 aerial photo analysis results of Site 19, former Camp Sibert. The green areas (around 3200 acres) are the calculated undisturbed areas, where geophysical survey mapping and intrusive investigations were mainly focused. The blue line represents the boundary of Site 19, with around 4900 acres. The brown lines represent the firing fans.

different values, to form patterns. Through analyzing similar and different patterns, features in an image can be identified and recognized. In Site 19, undisturbed areas are mainly covered with thick vegetation, while disturbed areas are usually bare surfaces, such as roads, farms, and developed properties. Their contrasting differences form the boundaries of the features.

During the image processing and analysis process, some errors were unavoidably introduced. Therefore, the extracted features from the aerial photos were then converted into vector data format using a GIS raster-to-vector conversion tool. They then could be further cleaned up and edited to better represent the true conditions of the project site.

The calculated total acreage of the undisturbed areas derived from the image analysis was around 3200 acres, around 65% of the whole 4900 acre site.

By focusing a geophysical survey and an intrusive investigation on these identified undisturbed areas instead of the whole site, savings of about one-third of the cost were achieved. The focused investigating areas were further reduced by overlaying them with firing fans, also shown in Figure 3.2. Since the areas within the firing fans were more likely to be impacted than the areas outside the firing fans, investigations were more focused on them. Other areas, less likely to be impacted, were also sampled, selectively surveyed, and intrusively investigated to ensure a sufficient amount of data was collected throughout the whole project site.

Case Study 3.3 (Site 2A of Former Camp Sibert) and 3.4 (Tanapag Fuel Farm): Analyzing Historical Aerial Photos and Drawings

In 2009, the Chemical Agent Contaminated Media Removal Action (RA) and the Remedial Investigation/Feasibility Study (RI/FS) projects were conducted at Site 2A, which is situated in the central portion of the former Camp Sibert. Site 2A was formerly used for chemical agent decontamination training and also as a burial site for training materials. Training at Site 2A was mainly decontamination training on an airplane fuselage as well as on walls, floors, different road types (gravel, concrete, sand and macadam), shell holes, and trucks. In addition to the burial pits, the site also contained a mustard soakage pit, a chemical agent storage area, a supply building and a shower/dressing facility. Historical records and previous investigation reports indicate that the chemical agents used for training or buried at Site 2A include mustard, nitrogen-mustards, and lewisite. Industrial chemicals such as phosgene, fuming sulfuric acid, tearing agent and adamsite might also have been used. At the end of World War II in late 1945, training activities at former Camp Sibert ended and the whole site was closed. CWM-related materials and equipment left over from the training were buried at Site 2A in three burial pits and excess chemical agents were poured into a soakage pit.

GIS image analysis techniques were used to locate and delineate the decontamination training areas, burial pits, mustard soakage pit, chemical agent storage area, supply building and shower/dressing facility. Historical drawings, showing the training and burial layout of Site 2A, were scanned, geo-referenced, and overlaid on a recent aerial photo to identify these impacted areas, as shown in Figure 3.3. Historical aerial photos were also processed and analyzed to confirm the impacted areas identified from the drawing and to further refine their locations and boundaries.

These kinds of information were very useful for the project team and their field crews in selecting affected properties and obtaining ROE; planning brush clearing; constructing site access roads; designing geophysical survey areas; and identifying soil, surface water, sediment and groundwater sampling locations. This information also fed into the work to identify intrusive investigation approaches and to calculate safety zones, as shown in Figure 3.4.

FIGURE 3.3

A 1944 historical drawing showing the layout of Site 2A. It was scanned, georeferenced, and overlaid on a recent aerial photo to locate and delineate the burial pits, mustard soakage pit, chemical agent storage area, supply building, shower/dressing facility and the decontamination training areas for the airplane fuselage, walls, floors, roads, shell holes, and trucks. The red lines represent project sites. The middle one is Site 2A. The blue line polygons are the identified impact areas, including three CWM burial pits in the north of Site 2A and a mustard soakage pit, a chemical agent storage area, a supply building and a shower/dressing facility, truck decontamination area and shell holes decontamination area in the south of Site 2A.

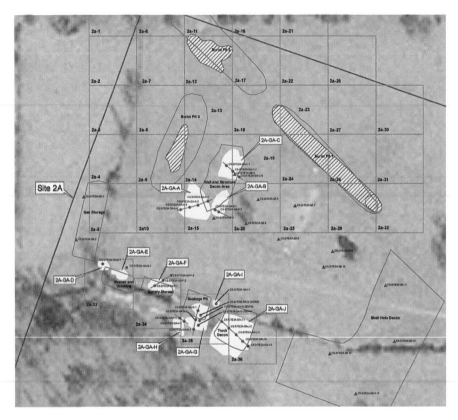

FIGURE 3.4

A 1944 historical aerial photo of Site 2A, showing the features related to the CWM decontamination training and the burial activities at the site from 1942 to late 1945. The red lines represent project sites. The one in the center of the map is Site 2A. The blue line polygons are the identified impact areas, including three CWM burial pits in the north of Site 2A and a mustard soakage pit, a chemical agent storage area, a supply building and a shower/dressing facility, truck decontamination area and shell holes decontamination area in the south of Site 2A. Samples (red dots and red triangles), trenches (lines), geophysical survey (yellow areas), and others were planned to survey and investigate these identified impact areas.

The next case study (i.e., case study 3.4) discusses how historical aerial photos were processed and analyzed to locate fuel storage tanks in the Tanapag Fuel Farm (TFF), a formerly used fuel storage facility (also known as a fuel farm) in Tanapag village in the northwestern section of Saipan Island. Saipan is the largest island of the Commonwealth of the Northern Mariana Islands (CNMI), a north–south trending chain containing 16 small islands in the tropical Western Pacific. Saipan, with an area of about 48 square miles and a population of approximately 65,000, is the center of government, transportation, commerce, and education for the CNMI.

The former TFF actually contains two separate sites, the northeast site and the southwest site, which lie approximately 0.6 mile apart at their closest points. The smaller, 4.8 acre southwest site is located in the region known as Puerto Rico, while the larger, 96 acre northeast site is located in the village of Tanapag. Classified as a tropical marine climate zone, Saipan is warm and humid throughout the year, with an average temperature of around 75°–80°F and a mean annual rainfall of 80 in.

Historically, Saipan was under the control of Japan for some time, until U.S. forces invaded it in June 1944, during World War II. The fuel farm was built by U.S. forces to supply fuel oil, diesel fuel and aviation gasoline for U.S. Navy ships and aircraft during World War II and through the 1950s, after which the fuel storage tanks were abandoned. Declassified records and previous reports indicated that between 40 and 47 above-ground tanks were constructed at the two TFF sites, with the majority of the tanks located at the northeast site and only a few at the southwest site. The tanks fall into three broad categories: the 10,000 barrel (bbl) tanks for special Navy fuel oil; 10,000 bbl tanks for diesel fuel; and 1,000 bbl tanks for aviation gas. In addition to the fuel storage tanks, there were numerous connecting pipelines, pump houses and 1.2 miles of submarine lines for each type of fuel.

Due to both human activity and the elements, these abandoned fuel storage tanks were partially or completely collapsed, rusted, corroded, crushed, or simply left as piles of metal debris. They became physical and chemical hazards to human health and the environment. Chemical hazards include total petroleum hydrocarbons (oil) as the primary contaminant of concern and also some metals such as arsenic, cadmium, chromium and iron.

The primary objectives of the proposed RA project included a biological survey for endangered species; brush clearance and access road improvements; demolition and recycling of around 17 tanks (including disposal of sludge and residual fuel); sampling and characterization of wastes for disposal; and an environmental investigation of impacts on soil, surface water, and groundwater related to the tanks.

During review of historical records and previous investigation reports, discrepancies were found regarding the total number of the above-ground fuel storage tanks and their locations. It was suspected that the locations of some of the tanks were significantly inaccurate. Therefore, GIS image analysis tools were used to process and analyze historical aerial photos of the mid- to late 1940s and 1950s to accurately locate the tanks. Figure 3.5 is an analyzed 1946 high-resolution aerial photo map (top) showing the tank locations at the northeast site of the TFF, while the bottom is a picture of a collapsed tank. Other historical aerial photos were also processed and analyzed to confirm the tank locations in both the northeast and southwest sites.

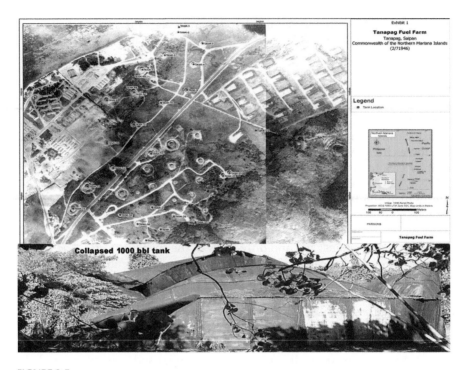

FIGURE 3.5
The top map is a georeferenced 1946 high-resolution aerial photo showing fuel storage tanks in the northeast Tanapag Fuel Farm site in Saipan, the Commonwealth of the Northern Mariana Islands. The bottom is a picture of a collapsed fuel storage tank. The red points represent the fuel storage tanks selected for the proposed Removal Action project.

Case Study 3.5: Analyzing High-Resolution LiDAR Data

A large (approximately 12,831 acres) formerly used defense site (FUDS) in central Louisiana state was used for air-to-ground bombing and small arms training in the 1940s. After the training ended and the FUDS closed in the early 1950s, the property was transferred to the U.S. National Forest Service (NFS). There were several reports about OE items encountered by NFS employees in the FUDS. In 2005, a site inspections project was conducted at this munitions and explosives of concern (MEC) impacted site. Since the entire large site was covered with heavy vegetation, it was difficult to locate the MEC-impacted areas because the features brought about by the historical bombing and firing related activities were not visible on the ground. Not enough historical drawings or aerial photos could be found to identify the impacted areas either. Also, since the site is so large and located in a national forest, it was virtually impossible to clear the vegetation from the whole site. Therefore, light detection and ranging (LiDAR) technology, which can penetrate thick vegetation, was used to survey the site to locate the features and delineate the impacted areas.

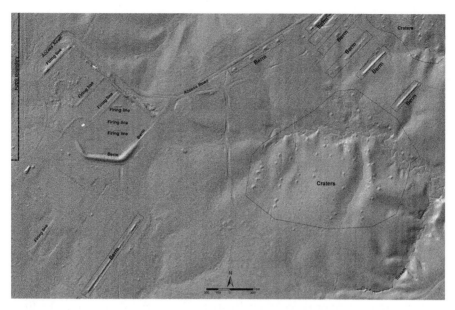

FIGURE 3.6
High-resolution LiDAR images showing the MEC impacted areas in a formerly used defense site (FUDS) in central Louisiana.

LiDAR is a remote sensing technology that uses light in the form of a pulsed laser to measure distances to the Earth. The light pulses, which can penetrate vegetation, combined with other data recorded by the airborne system, can generate high-resolution 3D information about the shape of the Earth's surface. The LiDAR survey data were processed with GIS image analysis techniques to produce high-resolution 3D images of the MEC-impacted site, as shown in Figure 3.6. The processed LiDAR images clearly show the features from the historical training activities, such as bombing craters, firing lines, berms, site access roads and so on. With these features identified, the MEC-impacted areas were located and their boundaries were accurately delineated. Site investigation approaches were designed to focus on the MEC-related areas to minimize impacts to the habitats of the protected species, wetlands, soil, and plants in the national forest.

Case Study 3.6: Topographic Slope Analysis

Slope (also known as grade, gradient, incline, pitch or rise) information is very important in civil engineering, environmental investigations, geology, hydrology, transportation, agriculture, mining, conservation, infrastructure and many other applications. In mathematics, a slope is defined as the ratio of the vertical change to the horizontal change of a surface. The vertical change is called the rise, while the horizontal change is called the run.

There are two different ways to express a slope. One is the angle of a slope in the range of 0°–90°, and the second is to assign it a percentage gradient, also known as a percent rise. The degree of a slope is calculated as the arctangent (also commonly known as arctan, atan, or tan⁻¹) of the angle of the inclined surface to the horizontal surface (i.e., *degree of the slope = arctan (rise/run)*). The percent of a slope is calculated as the rise divided by the run and multiplied by 100 (i.e., *percent of the slope = (rise/run) * 100*).

For example, when a slope's angle is at 45°, the rise is equal to the run. Therefore, the percent of the slope is 100%. When the slope angle decreases to 0°, the percent of the slope also becomes 0%, because the rise is 0. When the slope angle approaches 90°, that is, near vertical, the run approaches 0 and therefore the percent of the slope begins to approach infinity. Therefore, the percentage range of the slope is from 0 to near infinity. In other words, a flat surface is 0%, a 45° slope is 100%, and as the slope becomes steeper toward vertical, the percent gradient increases toward infinity. Regardless of the slope unit (angle or percent gradient), the lower the slope value, the flatter the terrain surface is. The higher the slope value, the steeper the terrain is. The conversion between these two slope types is *degree of the slope = arctan (percent of the slope/100)*.

An inclined surface has two properties: its slope (the maximum rate of change of the elevation of the surface) and its aspect (the facing direction of the surface with respect to the north). Aspect is measured clockwise in degrees from 0 (due north) to 360 (again due north) forming a full circle. With respect to its facing direction or aspect, the slope value of a surface can be either positive or negative. An upward surface has a positive slope value, while a downward surface has a negative slope value. A flat surface has a zero slope.

For environmental investigations, especially ones requiring walking or driving around the site for visual reconnaissance and geophysical survey mapping, topographic slope information is needed to design visual reconnaissance paths (also known as qualitative reconnaissance tracks) and geophysical survey paths (also known as transects). Some geophysical survey equipment, such as carried, cart-mounted, or towed-array geophysical systems, cannot be used in areas with steep slopes, for example, those near or above 30°. Also, for a site contaminated by ordnance and explosive waste, investigations should be focused more on downslope areas and less on upslope areas, because MECs, MDs, and MCs are usually found on the surface or relatively shallow in the ground and therefore move down on slopes.

This case study is the FUDS Military Munitions Response Program (MMRP) Remedial Investigation/Feasibility Study (RI/FS) of the former Waikoloa Maneuver Area on the Island of Hawaii. In December 1943, the U.S. Navy acquired approximately 91,000 acres of land from a private ranch. Portions of the land were used as an artillery firing range, while other portions were used for troop maneuvers. Intensive live-fire training was conducted in forested areas, cane fields, and around the hills. Weapons used in the training exercises include carbines, rifles, bazookas, flame-throwers, hand grenades, mortars and machine guns, and so on. They also test-fired packages of Japanese language surrender leaflets. Infantry regiments conducted maneuvers with fighter and dive-bomber support. There was also other training with demolitions, mines, and other special equipment, such as construction equipment (e.g., cranes, bulldozers, and tractors), tanks, weapon carriers, trucks and jeeps. After World War II, the training site was closed.

The current and future land uses are mainly residential, agricultural, pastoral, conservational, industrial, and resort/recreational uses. Topographically, it is divided into two landforms, the steep sloped upland areas and the gentle sloped lava plains. The upland is cut by widely spaced gullies formed by erosion.

For this large project site, many 7.5-minute quadrangles of the U.S. Geological Survey (USGS) Digital Elevation Model (DEM) data and some higher-resolution (1 meter) National Elevation Dataset (NED) data were obtained and used for slope analysis, depending on data availability of the areas.

The DEM is a more commonly used generic term for digital elevation data. There are other terms and formats of this kind of elevation data, such as digital terrain model (DTM) and digital surface model (DSM). In the 1970s, the USGS started producing DEMs from aerial imagery using manual stereoplotters. Since then, numerous methods have been developed to produce DEM data more efficiently and accurately, especially by remote sensing technologies, such as the SPOT 1 satellite, the European Remote Sensing Satellite (ERS), the Shuttle Radar Topography Mission (SRTM), and others. The major DEM data providers include USGS, CGIAR, Spot Image, and the Earth Remote Sensing Data Analysis Center (ERSDAC, now 'Japan Space Systems'). Starting in early 1990s, the USGS has been producing seamless elevation data, named National Elevation Datasets (NED), covering the United States, Hawaii, and Puerto Rico from its older DEMs and other sources.

GIS spatial analysis tools were utilized to process and analyze USGS DEM and NED datasets to calculate slopes at this large MMRP site. In GIS, for each raster cell, its slope is derived by calculating the maximum rate of change in elevation value from the subject cell to its eight neighboring cells in a 3×3 cells window. The subject cell is at the center of the 3×3 cells window. By moving this window, the slope of every cell in the elevation raster dataset is calculated.

Figure 3.7 is an example slope analysis results map of Project 20 (also named as Sector 16) located in the northern portion of the former Waikoloa Maneuver Area (see the Index Map of Figure 3.8). The northern part of Project 20 is the steep sloped upland in the foothills of the Kohala Mountains with some four-wheel drive (4WD) trails (dashed lines

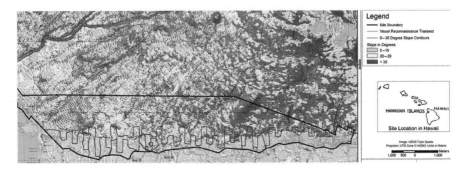

FIGURE 3.7
Slope analysis map of Project 20 located in the northern portion of the former Waikoloa Maneuver Area. The red areas represent steep slopes (>30°), the yellow areas representing moderately steep slopes (20°–29°) and the green areas representing a gentle slope (<19°). This slope analysis map is useful in designing field work activities, such as qualitative reconnaissance tracks, geophysical transects and intrusive investigations, as shown on Figure 3.8.

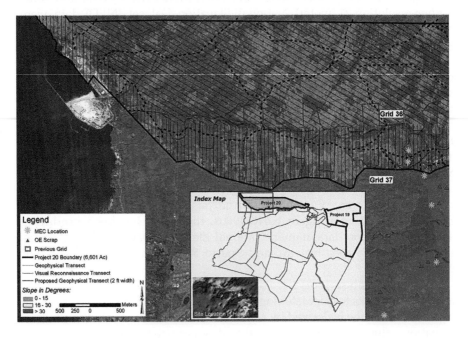

FIGURE 3.8
Proposed field work map of Project 20 of the former Waikoloa Maneuver Area based on the slope analysis results. The spacing and orientation of the qualitative reconnaissance tracks (purple lines) and geophysical transects (black lines) were designed based on the slope analysis of the project site. The red areas represent steep slopes (>30°), the yellow areas represent moderately steep slopes (20°–29°), and the green areas representing gentle slopes (<19°). Qualitative reconnaissance, geophysical survey and intrusive investigations are more focused in the downslope areas as compared with the steep upslope areas. Also, proposed qualitative reconnaissance tracks and geophysical survey transects follow the general topographic elevation contour lines to minimize climbing the steep slopes.

in Figure 3.8). The southern part of Project 20 is a strip of gently sloped lava plain.

Due to these contrasting topographic characteristics, Project 20 is divided into two investigation sections with different approaches. As displayed in Figure 3.8, investigation will be more focused on the relatively flat area in the south, with dense planned qualitative reconnaissance tracks, geophysical transects, and intrusive investigation grids. In the northern part, with challenging steep sloped terrains, relatively less transects are planned. Also, planned transect lines are oriented following the general slope patterns of the terrains (i.e., along elevation contours) to avoid steep climbing against the slopes.

Case Study 3.7: Geophysical Anomaly Density Analysis

A remedial investigation and feasibility study (RI/FS) project and RA project were conducted at seven conventional MEC-impacted sites and four suspected CWM-impacted sites inside former Camp Sibert in 2010 and 2011 (as shown in Figure 3.9). The primary goal of the remedial investigation (RI)was to collect a sufficient amount of data to determine the nature and extent of MEC, CWM and MC pollution. The data and results from the RI were then used by the Feasibility Study phase of the project to develop and evaluate effective remedial alternatives. During the RI, datasets collected from the digital geophysical mapping (DGM), the intrusive investigation of geophysical anomalies, and the mag-and-dig investigation were used to assess the nature and extent of MEC and MC. The term "mag-and-dig" refers to an investigation method of using an analog geophysical instrument to locate an anomaly, marking the detected anomaly with a flag and excavating the flagged location to determine the item(s) buried there.

During the project's FS phase, different remedial alternatives were developed and assessed for managing risks associated with potential MEC and MC contamination. Its main purpose was to provide decision-makers with the necessary data to further evaluate and select the final remedies for the sites. The analysis of the density of MEC and MD is one of the most important factors in developing and evaluating the remedial alternatives for a site.

A large volume of the datasets from a variety of previous investigations and the current RI project were compiled, processed, and statistically analyzed to generate the MEC/MD density maps for the whole former Camp Sibert, as well as the individual sites selected for these RI/FS and RA projects.

Basically, density is the result of quantity divided by area, that is, the magnitude of something per square unit. In statistics, density analysis is a type

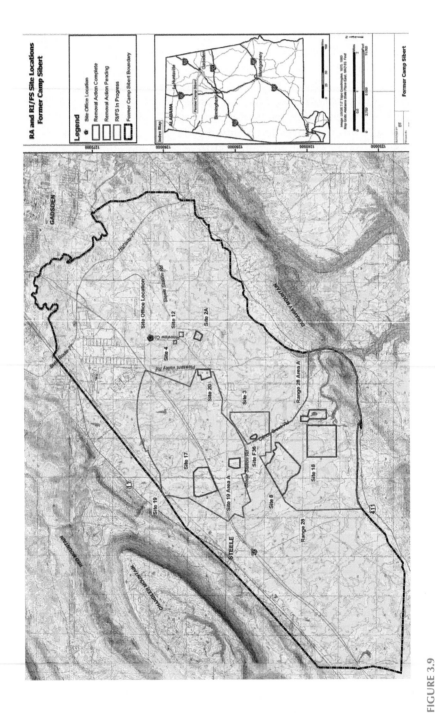

FIGURE 3.9
A GIS map showing RA and RI/FS project sites inside the former Camp Sibert. Some of the sites are conventional Munitions and Explosives of Concern (MEC) impacted sites, while others are suspected chemical warfare materiel (CWM) impacted sites, or both MEC and CWM.

of spatial interpolation, which analyzes the spatial relationships, patterns, and trends of the measured/known quantities of samples and spreads them across the whole study area based on a data interpolation method. Densities of both point and line features can be calculated. Line density calculates the density of linear features in the neighborhood of each output raster cell, in the unit of length per square unit of the area of interest, for example, the density of streams in a watershed study area, the density of roads in a proposed distribution warehouse area and the density of fractures and/or faults in an engineering project area.

However, point density information is more commonly used in research, studies, planning, and others, such as population density of a region, crime density of a neighborhood or city, accident density of a traffic study area, geophysical anomaly density of an environmental project area, and many more.

There are two density analysis methods in GIS: point density and line density, with simple and kernel options. For a point density analysis with the simple option, the study area is divided into a raster grid, using the cell size input value. Then, a search circle is drawn around each raster cell center using the search radius input value. All the points within the search circle are multiplied by its population (also known as weight) field value in the point feature's attribute table, summed, and divided by the circle's area to get the density of the raster cell. The population field value assigns weight to the data point location. It is basically the quantity at that spatial location, for example, the number of people at an address, the number of crimes recorded at a location over a certain period of time, or the number of geophysical anomalies discovered in a survey location. The smaller the cell size, the higher the output data resolution is. The larger the search radius, the larger the search circle is. Therefore, more data points are included in each cell calculation. It is important to understand the characteristics of the data points and the study area to set the appropriate cell size and search radius to achieve more accurate and meaningful density calculation results.

Similar to the point density method, line density analysis measures the length of the portion of each line that falls within the search circle, multiplies the length by its population field value in the line feature's attribute table, adds them together, and divides the total by the circle's area to get the line density of the raster cell.

What the kernel option does in a point density analysis process is fit a smoothly curved surface over each data point location where the surface value is the highest and then decreases gradually with increasing distance away from the data point until reaching zero at end of the data point's search radius. The density of each grid cell is calculated by adding the values of all the kernel surfaces where they overlay the grid cell center. The kernel option usually produces a smoother density surface over a large area than the simple option. However, in a large area with data clusters and big data gaps between them, such as the MEC/MD data points shown in Figure 3.10, the kernel option could produce an inaccurately smooth density surface over the study area by overstretching/smoothing the data points to fill the data gaps.

If not specified, density analysis in GIS will use the extent of the input data layer as the boundary of the calculation. Therefore, points on the edges of the data layer will have truncated densities. However, in most cases, it is usually desirable to have density calculated slightly beyond the data points. This can be achieved by selecting a slightly larger polygon layer as the extent of the density calculation, in the "environments" setting of the density analysis process.

Figure 3.10 shows the MEC/MD density patterns of multiple sites inside the former Camp Sibert. It was calculated using the point density method with the simple option, using the project site boundaries as the extents. A variety of cell sizes and search radius values were experimented with to achieve the most accurate, reasonable, and meaningful density results, which were evaluated based on the datasets and knowledge of the conditions of the project sites. Figure 3.11 is a MEC/MD density map of Site 18, which is located in the southeastern section of the former Camp Sibert.

The MEC/MD density analysis results and other relevant data were used in the FS to develop and screen the seven remedial alternatives, which are briefly explained as follows:

1. No Further Action (NFA): No action is needed for the site.

2. Educational Awareness: A site-specific educational awareness program is needed, such as fact sheets, public involvement, and so on.

3. Surface MEC Removal with MC Contaminated Soil Removal at UXO Detonation Locations: Only MEC items on surface and surface soil contaminated by munitions constituents at UXO detonation locations are required to be removed.

4. MEC Removal to 1 Foot Depth with MC Contaminated Soil Removal at UXO Detonation Locations: MEC items and soil up to a depth of 1 foot at UXO detonation locations are required to be removed.

5. MEC Removal to Maximum Depth of UXO or MD with MC Contaminated Soil Removal at UXO Detonation Locations: An MEC removal will be conducted to the maximum depth of the UXO or MD recovered from the specific MRS during the RI or previous investigations along with MC contaminated soil removal at UXO detonation locations.

6. Fencing and Signs: Installing fencing and signs around the site to restrict access and minimize possible receptor interaction.

7. Excavation, Sifting, and Restoration: Completely remove all MEC and munitions debris items from the site and restore it.

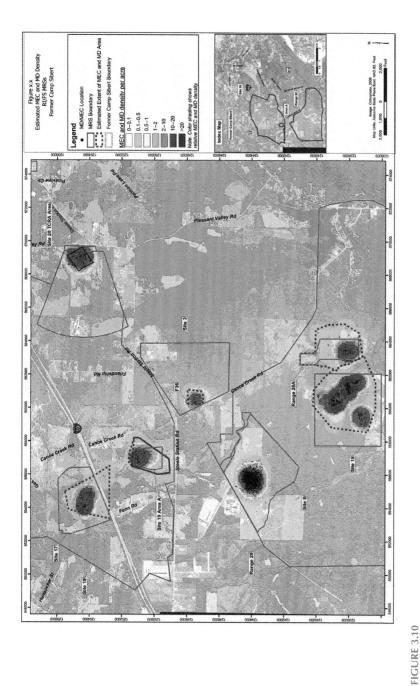

FIGURE 3.10

MEC/MD density map of the RA and RI/FS sites inside the former Camp Sibert. The blue lines are the boundaries of the RA or RI/FS project sites, also known as a munitions response site (MRS). The orange dashed line is the former Camp Sibert boundary. The dark red dashed lines are the boundaries of the estimated higher MEC and MD areas. The small black points are the MEC/MD locations. Red represents the highest MEC/MD density area, followed by pink, orange, yellow, and green, representing lower MEC/MD density areas. The MEC/MD density unit is density per acre.

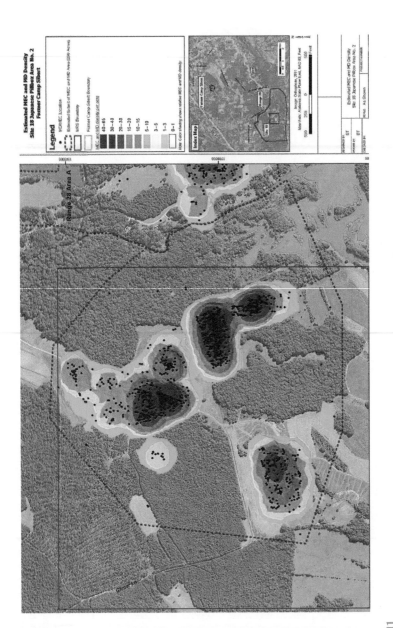

FIGURE 3.11

An example MEC/MD density map of Site 18, which is a conventional Munitions and Explosives of Concern (MEC) impacted site. Similar MEC/MD density maps were generated for all the selected Removal Action (RA) and Remedial Investigation and Feasibility Study (RI/FS) sites inside the former Camp Sibert. The red line represents the boundary of the RA or RI/FS project site, also known as munitions response site (MRS). The orange dashed line is the former Camp Sibert boundary. Red dashed lines are the boundaries of the estimated higher MEC and MD areas. The small black points are the MEC/MD locations. Red represents the highest MEC/MD density area, followed by pink, purple, orange, yellow, and green, representing lower MEC/MD density areas. The MEC/MD density unit is density per acre.

Case Study 3.8: Chemical Concentration Analysis

During the 2009 chemical agent contaminated media RA and the RI/FS investigations at Site 2A of the former Camp Sibert, the contaminated soil was sampled for the presence and levels of contamination from chemical agents and agent breakdown products. As discussed earlier in this chapter, Site 2A was formerly used for chemical agent decontamination training between 1942 and 1945 during World War II. When the war ended in late 1945, the chemical training at Site 2A was terminated and excess chemical agents were poured into the mustard soakage pit. Many soil samples were taken from the mustard soakage pit, and arsenic was detected in all of the samples at concentrations ranging from 640J µg/kg up to 140,000 µg/kg. Arsenic in 20 of those samples exceeded the established background level of 7,659 µg/kg (~8 mg/kg). The highest detected arsenic concentration was in a floor sample (SB-35) collected at a depth of 4 feet, as shown in Table 3.1. It is an example small data table queried out from a chemical analytical results database for illustration purposes.

This example table contains valuable information about the arsenic contamination of the mustard soakage pit. However, by simply looking at the numbers in the table alone, it is hard to understand and visualize the whole contamination situation of the area both on the surface and underground. In order to visualize the spatial distribution of the arsenic concentrations, this data table was joined to the sampling locations GIS data layer to generate a tag map showing both sample locations on the ground surface and their associated arsenic analytical results at all the sampled depth intervals underground, as shown in Figure 3.12. A GIS program was written to query out the arsenic results of each sample location, sort the arsenic results based on their sampling intervals, and automatically create a mini data table (also called a tag) next to the sample location. For a site with many samples and large volumes of chemical analytical data, this automated mapping process saved not only the time taken to generate a large number of mini data tables (or tags), which is tedious and error-prone cartographic work, but also avoids the efforts to QA/QC (quality assurance/quality control) the chemical numbers inside the tags, because no possible human data entry errors are introduced in the automated data processing and mapping work flow. Therefore, only the source data tables in the database need to be reviewed and checked. If the source data tables in the database are correct and accurate, then all the GIS tag maps automatically generated from the data tables should be free of errors. When the source data tables are updated, the GIS tag maps are also automatically updated. Also, the GIS program can be modified to display other types of data or measurements onto maps. It significantly increases mapping efficiency and data quality and reduces data QA/QC time and effort.

TABLE 3.1

An Example Small Portion of a Large Arsenic Concentration (mg/kg) Data Table Queried Out from a Master Database of the Mustard Soakage Pit inside Site 2A of the Former Camp Sibert

Depth	SB 31	SB 32	SB 33	SB 34	SB 35	SB 36	SB 37	SB 38	SB 45	SB 46	SB 47	SB 48	SB 49	SB 50	SB 51
					Mustard Soakage Pit Preliminary Arsenic Results in mg/kg										
0–2	100	16	4.3	3.4	380	67	2.8	60	3.2	140	3.7	5.1	4.2		16
2–4	n/a	n/a	n/a	n/a	750	80	2.3	6.1	2.7	154	3.4	5.6	4.9		15
3–5	720	2.8	2	3	n/a	n/a	n/a	n/a	n/a	n/a	n/a	n/a	n/a	n/a	n/a
4–6	n/a	n/a	n/a	n/a	1500	1.9	3.1	10	1.4		3.3	16.1	5.1		73
5–7	n/a	n/a	n/a	3.6	n/a	n/a	n/a	n/a	n/a	n/a	n/a	n/a	n/a	n/a	n/a
6–8	n/a	n/a	n/a	n/a	150	2.7	2.1	50	1.5	2.3	4.9	2.8	37.6		9.5
8–10	n/a	n/a	n/a	n/a	77	2.2	2.1	9.2	3.9		3.8	2.7	5		7.6
10–12	n/a	n/a	n/a	n/a	93	1.9	1.5	4.1	1.9	2.5	3.6	3.3	3.4	2.9	
12–14	n/a	n/a	n/a	n/a	95	2.1	2.4	3.4	3	4.6	3	3.7	21.6	2.6	320
14–16	n/a	n/a	n/a	n/a	33	2.6	2.5	18	1.8		2.2	33.8	2	6.3	7.7
16–18	n/a	n/a	n/a	n/a	51	2	1.6	160	1.5	67	2.5	7.2	ND	3.4	3.2
18–20	n/a	n/a	n/a	n/a	3.2	n/a	2.3	8.2	1.7	11	1.545	2	2.6	3.8	1.3

Note: Many samples were collected from different sampling depths, ranging from 0 to 20 feet and analyzed.
1. Background is 7.8 mg/kg.
2. Bold concentrations are above background.

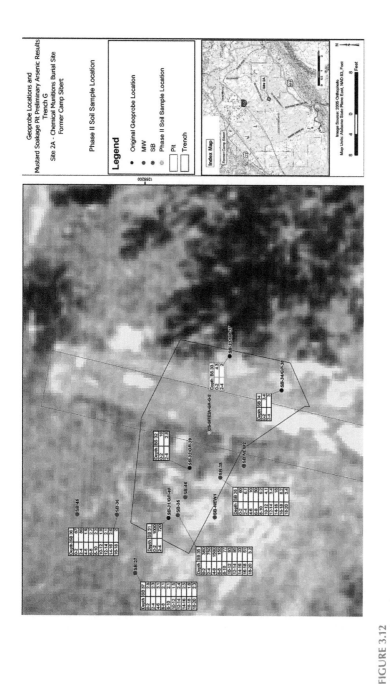

FIGURE 3.12

Arsenic analytical results tag map of the mustard soakage pit in Site 2A of the former Camp Sibert. The black line represents the mustard soakage pit boundary. The blue line represents an intrusive investigation trench. The red points represent soil boring/sampling locations. The black points are geoprobe locations. The green points represent Phase II soil sampling locations. A chemical analytical data table in a master database was joined to the sample GIS data layer to display both the spatial locations of the samples and their arsenic concentrations at different sampling depth intervals underground. The mini data tables (also known as tags) placed next the samples on this GIS map were generated automatically from the data in the master chemical database with a GIS program written for creating this type of mini table formatted special labels, also known as tags.

With this data tag map, as shown in Figure 3.12, it was much easier to see the arsenic concentrations geospatially rather than simply reading the data tables. However, with so many sampling depth intervals, it is still not easy to visualize arsenic concentration patterns at each internal depth and their relationships. To further help scientists and decision-makers to better understand the data and visualize site conditions, the arsenic concentration data from all the samples and at all depth intervals were statistically analyzed using GIS geospatial analysis techniques, based on Kriging, inverse distance weight, and other interpolation methods. The analyzed results were displayed as color-shaded concentration maps, one map for each depth interval. Figure 3.13 is an example color-shaded arsenic concentration contour map of sampling depth interval of 0–2 feet. Each contour map was carefully examined to ensure that the statistically analyzed results corresponded correctly and accurately with the actual data. If it did not, more statistical analysis experiments were performed with adjusted parameters and models until the

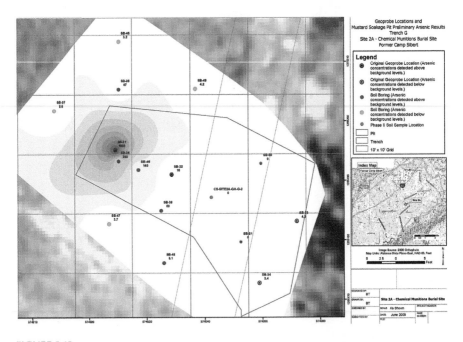

FIGURE 3.13
A color-shaded arsenic concentration (at 0–2 feet depth interval) contour map of the mustard soakage pit in Site 2A of the former Camp Sibert. The colored points are soil samples, labeled with their IDs (on top) and arsenic concentration values (on bottom). The bright red shaded areas represent soil with a higher arsenic concentration, while the light red shaded areas represent soil with a lower concentration level. The dark yellow squares represent the 10×10 feet intrusive investigation and soil removal grids. The boundary of the mustard soakage pit is represented by the pink line. The blue lines represent a trench excavated in the mustard soakage pit.

correct and accurate analytical results were achieved. With this type of color-shaded concentration contour maps, it is much easier to understand the arsenic concentration patterns at all the depth levels from 0 (surface) to 20 feet underground and estimate the soil volumes to be excavated and transported to the processing facility.

These geospatial analysis and visualization techniques added values to the existing datasets and revealed useful and meaningful information which was hidden in the large amount of numbers and which made it hard to see and understand. Three-dimensional modeling of the contaminated subsurface soil of this mustard soakage pit area will be discussed in more detail in Chapter 5.

Case Study 3.9: Analyzing Groundwater Elevations and Flow Directions with GIS

For an environmental project site with groundwater contaminations, investigation of its groundwater conditions, such as groundwater elevations and its flow patterns, is required to understand the sources of the contaminants, delineate a contaminated groundwater plume in 3D space, and predict its potential changes and movements. Tracking of temporal changes in contaminant concentrations is critical for the assessment of contaminated groundwater plume movement and its likely migration pathways and for designing and refining groundwater extraction and treatment systems to capture and remove the contaminants.

The formerly used Naval Air Station Brunswick (NASB) in the state of Maine is such an environmentally impacted site with a contaminated groundwater plume. The installation was originally constructed in 1943 mainly to train British Naval Command (Royal Canadian Air Force) pilots. During World War II, the facility also performed a secondary role of supporting anti-submarine warfare. After the war ended in late 1945, the installation was deactivated in October 1946. In 1949, the former NASB was used by the Brunswick Flying Service, a commercial aviation company. In 1951, during the Cold War, the station was re-commissioned as a naval air facility, mainly supporting three land-plane patrol squadrons, one fleet aircraft service squadron and a planned future mission as a master jet base. In December 1951, the naval a facility had its name officially changed to Naval Air Station Brunswick (NASB). The U.S. Air Force also used the installation for some missions from time to time. At the end of the Cold War in 1991, many operations at the naval air station were either reduced or relocated. In 2005, the installation was added to the Base Realignment and Closure list with a mandated closure date of September 2011. The former NASB site was officially decommissioned on May 31, 2011.

Due to the past operations, especially ordnance and fuel-related activities in the formerly used NASB, the soil and groundwater were contaminated in many areas, also known as sites. Figure 3.14 is a GIS map showing the layout of the formerly used naval air station and the environmentally impacted sites inside it. The elongated yellow area on the east side of the NASB is the Eastern Plume site. The groundwater under it was contaminated by dissolved-phase volatile organic compounds (VOCs). It is the focus of this case study. Some of the sites shown in Figure 3.14 were the sources of the contaminants.

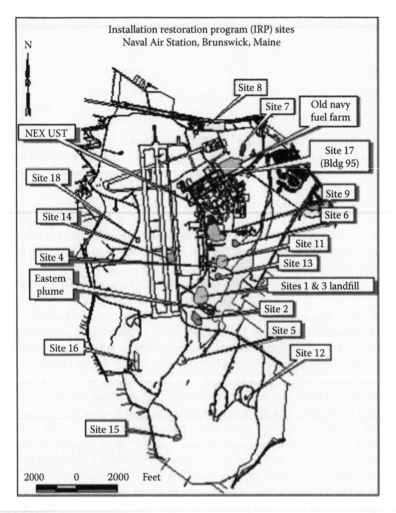

FIGURE 3.14

A GIS map showing environmentally impacted areas (also known as sites) in the formerly used Naval Air Station in Brunswick, Maine. The yellow elongated area is the Eastern Plume. The groundwater under it was contaminated by VOCs, forming a plume.

A long-term groundwater monitoring and sampling program was conducted from 1995 to the 2000s, with roughly two to four monitoring and sampling events per year. Long-term monitoring and sampling is widely used to assess the effectiveness of remedial approaches and monitor changes in contaminant concentrations in groundwater plumes over the time. The primary goal of this monitoring and sampling network was to monitor the changing conditions within the sand groundwater aquifer impacted by dissolved-phase VOCs including TCE, PCE, 1,1,1-TCA, and DCE in groundwater. There was an operating pump and treat system at the site of the Eastern Plume to remove groundwater contaminants and maintain hydraulic control of the VOCs-contaminated plume. The results of the monitoring and sampling were used to assess and improve the groundwater extraction and treatment system to make it more effective in capturing and removing contaminants.

Groundwater measurements and sampling were conducted at over 40 monitoring wells during many sampling events over seven years, resulting in the generation of more than 250,000 data records of chemical analytical data and groundwater measurements. GIS spatial analysis techniques were used in processing and analyzing these massive data records to model the contaminated groundwater plume in 2D and 3D spaces and to interpret groundwater elevation contours and flowing directions for each sampling event. Figure 3.15 contains groundwater elevation contours and flowing directions in the Eastern Plume area. The colored lines are groundwater elevation contours statistically analyzed from the groundwater measurement data from the long-term monitoring and sampling events. The blue lines are the groundwater elevation contours of the shallow aquifer and the red lines are the contours of the deep groundwater aquifer. The arrows are the interpreted groundwater flow directions from the contours. With the groundwater elevation data from all the long-term monitoring and sampling events in one map, it is much easier to visualize and understand the patterns of the groundwater conditions of the site, their changing trend during the monitoring events, the relationship between the source sites and the Eastern Plume, and the potential movement of the contaminated groundwater plume.

From groundwater measurement data, GIS can generate groundwater elevation contours automatically using Kriging, IDW, and many other statistical interpolation theories and models. Depending on the quantity and quality of the data, analysis results could vary largely.

Therefore, as with other types of statistical analysis, groundwater elevation contours generated automatically by computer should be reviewed carefully to ensure that they match the data correctly and accurately. If not, more experiments should be conducted with different statistical analysis methods and parameters until reasonably accurate results are achieved. It might also be necessary to manually edit computer generated contour lines to make them match the data and real conditions more closely. In this case, the automatically generated groundwater elevation contours GIS files were converted

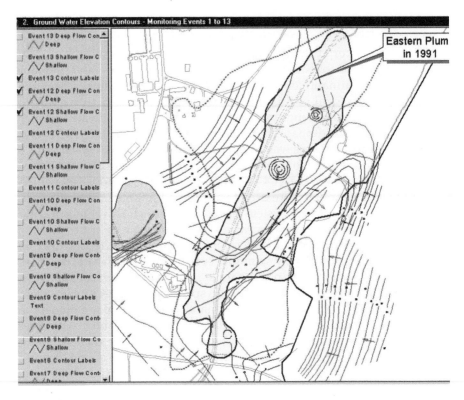

FIGURE 3.15

A GIS map showing groundwater elevation contours and flowing direction patterns and trends during the multiple sampling and monitoring events, in the Eastern Plume area, inside the formerly used Naval Air Station Brunswick, Maine. The colored lines are groundwater elevation contours generated from the groundwater measurement data from the long-term monitoring and sampling events. Blue lines are the groundwater elevation contours of the shallow aquifer, while the red lines represent the contours of the deep groundwater aquifer. The arrows represent the interpreted groundwater flow directions.

into AutoCAD drawings (in '.DXF' or '.DWG' format) for the knowledgeable geologists and environmental engineers to review and make edits. Then, the edited AutoCAD drawings were converted back into GIS data layers to be displayed and analyzed with other relevant datasets to investigate the sources of the groundwater contaminants, study the past and current sizes and conditions of the contaminated groundwater plume, model the plume's future changes and movements, and evaluate and enhance the pumping and treatment system to achieve the remediation goals.

Three-dimensional modeling of this contaminated groundwater plume will be discussed in more detail in Chapter 5. Three-dimensional modeling and visualization techniques would further enhance the understanding of the VOCs-contaminated groundwater plume and provide the information to make effective remedial decisions.

Case Study 3.10: Virgin Islands National Park General Management Plan/Environmental Impact Statement

The U.S. Virgin Islands are a group of islands in the Caribbean, including the larger islands of Saint Croix, Saint John, and Saint Thomas and many surrounding minor islands (Figure 3.16), with a total land area of around 133.7 square miles. They are geographically part of the Virgin Islands archipelago and are located in the Leeward Islands of the Lesser Antilles, which separates the Caribbean Sea from the Atlantic Ocean. The territory's capital is Charlotte Amalie on the island of Saint Thomas.

Virgin Islands National Park includes the majority of the island of St. John, as well as Hassel Island and Red Hook units on St. Thomas, with a total land area of approximately 7150 acres, as shown in Figure 3.17. Since the Virgin Islands National Park is mainly on St. John Island, the following discussion is focused on this portion of the park.

Virgin Islands National Park was established by Congress in December 1956 to administer and preserve the outstanding scenic and other features in

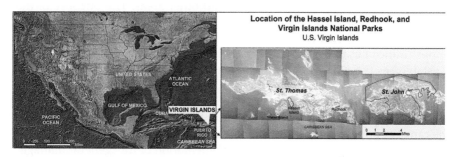

FIGURE 3.16
General location map of the U.S. Virgin Islands and the Virgin Islands National Park.

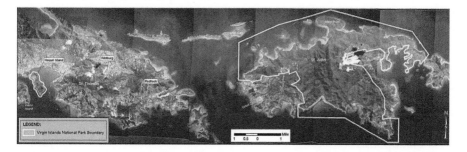

FIGURE 3.17
General layout of Virgin Islands National Park, which include the majority of St. John and Hassel Island and Red Hook in St. Thomas. The yellow line is the Virgin Islands National Park boundary.

a natural condition. The park is a nationally and internationally significant tropical environment where the processes of nature can be observed and studied. Due to its internationally significant natural resources, the Virgin Islands National Park was designated as a Biosphere Reserve by the United Nations Educational, Scientific, and Cultural Organization in 1976. The park contains a variety of terrestrial and marine species. Among them are many threatened and/or endangered species, such as the piping plover, Kirtland's warbler, roseate tern, green leatherback and hawksbill sea turtles, elkhorn and staghorn coral, the humpback whale, Thomas' lidflower, and St. Thomas prickly ash.

The park is also rich in cultural resources, with a variety of archeological sites dating from as early as 840 BC to the arrival of Columbus in the 1490s. Throughout the park, there are many historic landscapes and architectural remains of hundreds of structures from plantation estates. There are numerous ruins in the park, including windmills, animal mills, factories, great houses, terrace walls, and warehouses. Some historic structures, ruins, and sites are listed on the National Register of Historic Places.

Due to the significant growth in tourism and permanent/seasonal residents in the past five decades, a significant amount of development and construction has happened in St. John and other areas. The purpose of the general management plan is to provide comprehensive guidance for perpetuating natural systems, preserving cultural resources and providing opportunities for quality visitor experiences at Virgin Island National Park. It establishes the management framework for the park, addresses changing issues and conditions, incorporates new resource information, and provides management direction for new park lands.

The main purpose of the environmental impact statement is to identify the range of potential natural and cultural resources and environmental elements that could be affected by the implementation of this general management plan. A variety of relevant impact topics are selected and analyzed in the environmental impact statement, in accordance with the Council on Environmental Quality guidelines for implementing the National Environmental Policy Act and NPS management policies. The impact topics retained for analysis are:

- Air quality
- Soils
- Water resources (water quality and watershed conditions)
- Wetlands (mangroves and saltponds)
- Floodplains
- Vegetation
- Wildlife (marine and other wildlife species)
- Fish and marine invertebrates

- Marine resources (coral reefs, seagrasses, other types of bottom habitats, open water and other essential fish habitats)
- Special status species
- Soundscapes
- Scenic resources
- Cultural resources (archeological resources; historic structures, buildings and districts; cultural landscapes; ethnographic resources; collections and archival materials)
- Visitor use and experience (visitor use and access, recreational opportunity, and access to orientation information and interpretation)
- Socioeconomics
- Transportation
- Park operations and facilities (staffing, park facilities and maintenance, commercial services)
- Public health and safety
- Sustainability and long-term management

GIS technology was used extensively to collect and analyze a large amount of data in support of preparing the general management plan and the environmental impact statement. The datasets collected and analyzed include current and future land uses, real estate development, facilities, utilities, transportation, soil, wetlands, vegetation, water resources, habitats, cultural resources, census, and trails. Figure 3.18 shows the properties and subdivisions inside Virgin Islands National Park in St. John. There are many private lands and development inside the park. Figure 3.19 displays floodplains inside Virgin Islands National Park in St. John.

In order to better manage different areas of the park, four management zones were created, including the Visitor Contact and Operations Zone, the Recreation Zone, the Nature and Heritage Discovery Zone and the Resource Protection Zone, as shown in Figures 3.20 through 3.23. The focus of the Visitor Contact and Operations Zone is to provide access and support a wide variety of experiences and opportunities to obtain park information in a relatively developed setting. The Recreation Zone allows for a variety of experiences and opportunities including cultural and natural resource education with proximity to some facilities such as comfort stations, parking lots, and trails that provide access to areas where recreational opportunities are plentiful. The Nature and Heritage Discovery Zone represents areas that would provide access to and support a wide variety of educational opportunities to learn about the park's natural features and interpret the cultural heritage of hundreds of nationally recognized prehistoric and historic sites. The Resource Protection Zone focuses on resource preservation, protection, and scientific research and encompasses the core of the International Biosphere Reserve.

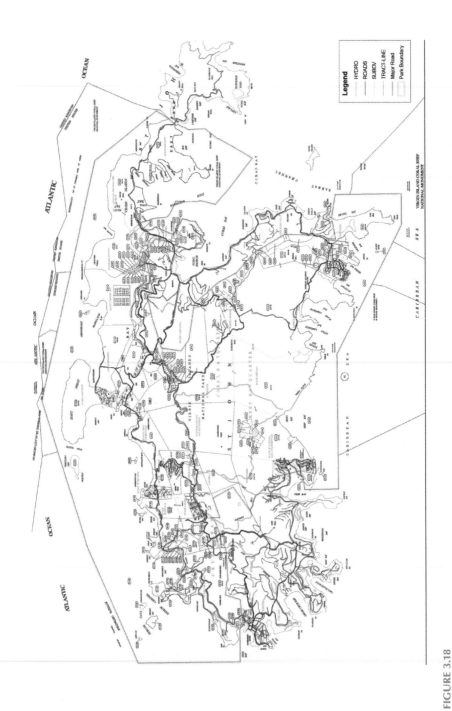

FIGURE 3.18
Properties and subdivisions inside Virgin Islands National Park in St. John. The yellow line is the Virgin Islands National Park boundary.

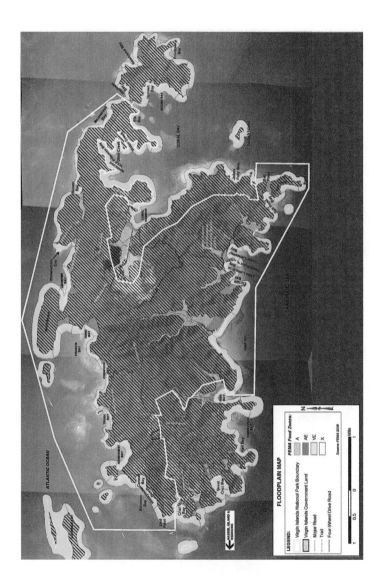

FIGURE 3.19

Floodplains inside Virgin Islands National Park in St. John. Floodplain data were obtained from Federal Emergency Management Agency (FEMA). Flood Zone A, high flood risk areas with a 1% annual chance of flooding and a 26% chance of flooding over the life of 30-year; Zone AE, high flood risk areas, similar to Zone A; Zone VE, high flood risk coastal areas with a 1% or greater chance of flooding and an additional hazard associated with storm waves and a 26% chance of flooding over the life of 30 years; Zone X, low flood risk areas, outside the 500-year flood and protected by levee from 100-year flood. The yellow line is the Virgin Islands National Park boundary.

FIGURE 3.20
Layout of Virgin Islands National Park in St. John, under Alternative A: the No Action Alternative.

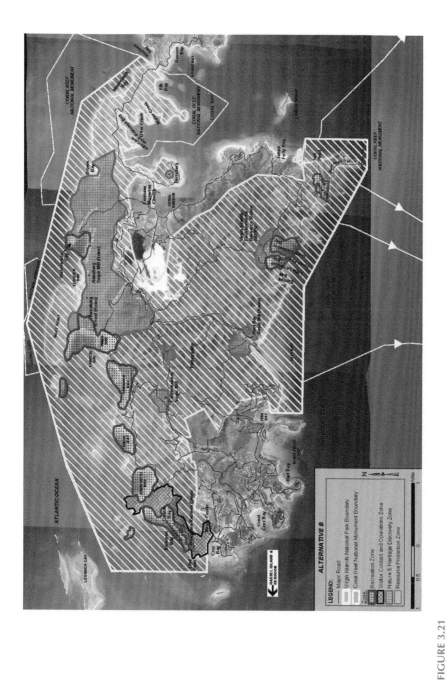

FIGURE 3.21
Layout of Virgin Islands National Park in St. John, under Alternative B: the Preferred Alternative.

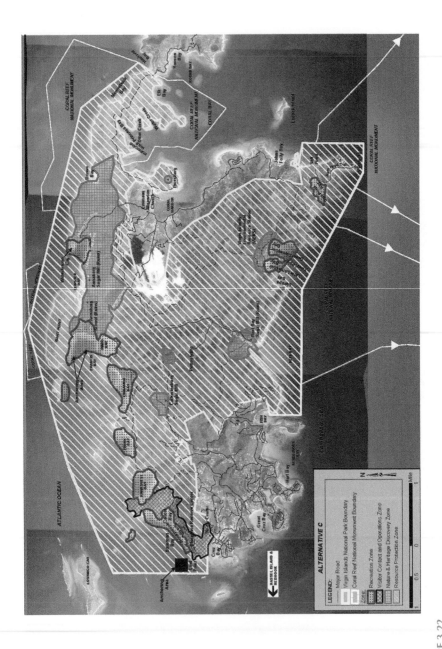

FIGURE 3.22
Layout of Virgin Islands National Park in St. John, under Alternative C.

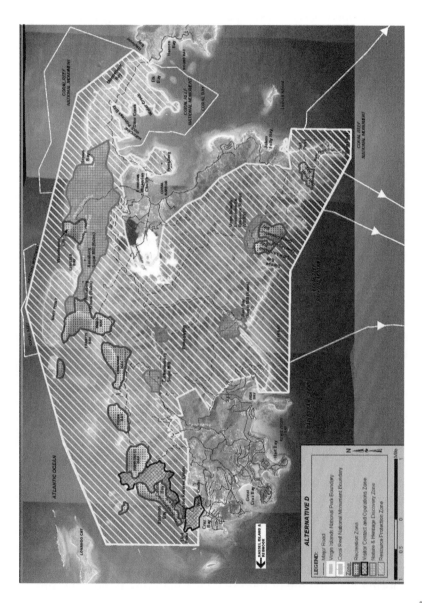

FIGURE 3.23

Layout of Virgin Islands National Park in St. John, under Alternative D.

After thoroughly collecting and analyzing the relevant data from the management zones and other areas, four alternatives were developed for managing visitor use and resources at Virgin Islands National Park, with each alternative providing a unique management approach. The alternatives were developed based on the park's purpose and significance, legal mandates, public views, and information on visitor use and park resources. The four alternatives are Alternative A, Alternative B, Alternative C, and Alternative D. They are discussed in detail in the following.

Alternative A

This alternative is also named the No Action alternative, that is, continuing current park management practices into the future. This No Action alternative serves as a baseline to evaluate the effects of the other three alternatives. It is also helps to explain why changes proposed in other alternatives for the future management of the park are necessary. Under this alternative, management practices, policies, and park programs, including maintenance of existing moorings, law enforcement, research, and operational practices, would continue without major changes. No new zones would be created under this No Action alternative either. The park would continue to be managed in accordance with the 1983 General Management Plan zones, as shown in Figure 3.20. They include the Natural Zone with Reef Subzone and Land Subzone, the Park Development Zone with Recreational Development Subzone, Access and Circulation Subzone, Administrative Development Subzone, a Residential Development Subzone, and the Historic Zone that includes key historic sites in the park.

Alternative B

Alternative B is the National Park Service Preferred Alternative. It was developed to address specific issues and concerns identified during the scoping process. The issues and concerns include

   I. Improving communication of scientific study results with the public through strategic, sustained, long-term, expanded education, outreach, and partnering efforts
  II. Adding new staff members to provide more consistent levels of education and enforcement activities
 III. Providing increased protection of natural resources by phasing out anchoring over the life of the 15–20 year planning period and providing approximately 15–23 new moorings
  IV. Implementing limited improvements of facilities and transportation systems and improving maintenance of facilities and historic resources

V. Applying adaptive management techniques for natural and cultural resources

VI. Implementing increased efforts to identify, stabilize, restore, and protect cultural resources

Compared with Alternative A (the No Action alternative), Alternative B enhances protective measures, education and visitor experiences at the park by establishing new management zones, which include the Resource Protection Zone, the Nature and Heritage Discovery Zone, the Visitor Contact and Operations Zone and the Recreation Zone, as illustrated in Figure 3.21. It also increases research, monitoring, enforcement, education, and partnering efforts.

Some new facilities are also proposed. They would be built using sustainable building practices in accordance with NPS policies and procedures to minimize footprints and disturbance of park resources. The NPS would participate in efforts to develop a jointly operated Museum/Environmental Heritage Center in Cruz Bay to provide a community center for cultural events, other special events, educational programs, and other activities.

Alternative C and Alternative D are similar to Alternative B, with only minor changes in the four management zones (i.e., Resource Protection Zone, Nature and Heritage Discovery Zone, Visitor Contact and Operations Zone and Recreation Zone), as shown in Figures 3.22 and 3.23 and other different management approaches and initiatives.

Executive Order 12898 requires all federal agencies to incorporate environmental justice into their missions by identifying and addressing disproportionately high and adverse human health or environmental effects of their programs and policies on minorities and low-income populations and communities. In compliance with this Executive Order and the National Environmental Policy Act, the proposed alternatives (A, B, C and D) were assessed during the planning process. It was determined that none of these four alternatives would result in disproportionately high, direct or indirect adverse effects on any minority or low-income populations or communities by analyzing the following information:

- The developments and actions in the alternatives would not result in any identifiable human health effects. Therefore, there would be no direct or indirect effects on human health within any minority or low-income population or community.
- The impacts on the natural and physical environment that would occur due to any of the alternatives would not disproportionately adversely affect any minority or low-income population or community, or be specific to such populations or communities.
- Impacts on the socioeconomic environment due to the implementation of actions proposed in the alternatives would be minor or

positive, and such impacts would not be expected to substantially alter the physical and social structure of nearby communities in St. John.

- The management alternatives would not cause hazardous materials to be generated or to affect the treatment of current hazardous materials.

All future development projects are subject to compliance with the National Environmental Policy Act, the National Historic Preservation Act, and other appropriate laws and regulations. Future project environmental reviews will be site specific and address natural and cultural resources, visitor experiences, and park operations.

4

Mapping and Analyzing Geology Data with GIS

GIS technology makes geologic data collecting, editing, analyzing, and mapping much easier than before. With mobile GIS technology, geologic data can be collected, edited, and mapped more efficiently and accurately in the field, in real time, similar to the environmental field data collecting and mapping processes discussed earlier. With desktop or server GIS applications, especially when they are programmed and customized, large quantities of geology data can be processed, analyzed, and compiled into geologic maps of any contents/themes, sizes, scales, and formats. Paper geologic maps produced in the past can be converted into digital GIS maps and data layers. Once in digital GIS format, geologic maps and their elements (i.e., data layers) can be manipulated, analyzed, and visualized in various ways to better understand complicated geologic units and structures. They can also be used to interpret their past and current relationships and transforming mechanisms on local, regional, and global scales, something that is virtually impossible by looking at paper geologic maps.

For example, the geologic data and maps collected and produced during the studies of the structural geology of southern California contributed to the understanding of the large-scale regional San Andreas Fault system, which is the most famous fault in the world. Extending around 810 miles along the coast throughout California, from the Salton Sea in the south to Cape Mendocino in the north, the San Andreas Fault system is the tectonic boundary between the Pacific Plate and the North American Plate. Its movements trigger earthquakes, such as the disastrous magnitude 7.8 San Francisco earthquake that occurred at 5:12 a.m. on April 18, 1906. The earthquake is one of the worst natural disasters in the history of the United States, damaging the majority of San Francisco and killing about 3000 people.

Case study 4.1: Geologic Mapping

In the past four to five decades, extensive research has been carried out on the structural geology of southern California and southwestern Arizona. Large amounts of field data—especially geologic structure measurements—have been collected, mainly by Professor Carl Jacobson of Iowa State University and his students, and Gordon Haxel, David Sherrod, and Richard Tosdal of the U.S. Geological Survey (USGS). Although geologists have tried to compile these geologic datasets onto maps and have them published, unfortunately, due to the high cost of the traditional map-production process, their hand-drawn maps and datasets were not officially published. Some of the geologic data were made available only as thesis maps or federal or state agency open-file reports. Based on their geologic survey data, they manually composed five 1:24,000 scale and one 1:50,000 scale geologic maps of the study areas in southwestern Arizona and southern California.

Figure 4.1 is a picture of the southern half portion of the 1:50,000 scale hand-drawn geologic map by the USGS geologists, Gordon Haxel, David Sherrod, and Richard Tosdal. The geologic features and symbols were hand-drawn on a Mylar sheet in the 1970s, based on the 7.5 minute USGS topographic quadrangles of Picacho, Picacho NW, Picacho SW, and Hidden

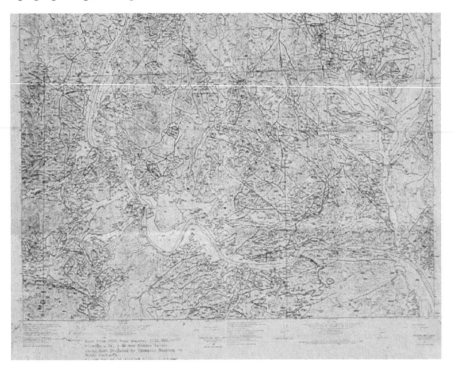

FIGURE 4.1
A picture of the southern half of the original hand-drawn 1:50,000 scale map entitled "Geologic Map of the Picacho, Picacho NW, Picacho SW, and Hidden Valley 7.5′ Quadrangles, Arizona and California" by D. R. Sherrod, R. M. Tosdal and G. B. Haxel. The geologic map was manually compiled on a Mylar sheet in 1970s by these geologists. With their permission, and funded by Professor C. Jacobson's National Science Foundation (NSF) grant, the author of this book converted this 1:50,000 scale and five other 1:24,000 scale hard copy geologic maps into digital GIS maps, with their associated GIS data layers, in 1997 and 1998 as part of his master's degree thesis.

Valley in southwestern Arizona and southern California. It contains many geologic units, structures, symbols and labels, as well as geographic, topographic, and hydrological features. Due to aging and intensively being used for research and teaching, the map was in poor condition, with fading lines, symbols, and labels. Although it is a valuable geologic map with a lot of important information and data, it is very hard to read and understand, especially for students and geologists not familiar with the general geology of the study area. It was hard to take a good quality picture of it as well due to its too-crowded features, symbols, small labels and poor map conditions.

I (the author of this book), as a graduate student of Professor Jacobson, participated in a geologic mapping and field data collection mission in southwestern Arizona and southern California in 1997. I was deeply impressed by the huge amount of geologic datasets collected by Professor Jacobson, his students, and the USGS geologists in the past years and the hand-drawn geologic maps on Mylar sheets, based on USGS 7.5 minute topographic quadrangles, as shown in Figure 4.1. Recognizing the scientific value of these unpublished geologic datasets and the hand-drawn paper maps, I decided to use GIS technology to convert the five 1:24,000 scale and one 1:50,000 scale hand-drawn hard copy geologic maps into digital GIS maps and plot the geologic measurement data onto the maps, so that both the geologic maps and GIS data layers could be easily accessed by more researchers, geologists, and students. Also, once in digital GIS format, the geologic maps and data layers could be manipulated, analyzed, and visualized to better understand the geology of the local study areas and the regional geologic structures of the western United States. The research and experiments on using GIS technology in geologic mapping, data processing, and analysis were part of my master's degree thesis, and were supported financially by Professor Jacobson's National Science Foundation (NSF) grant(s). The detailed documentation of the mapping and data producing processes, as well as the GIS and other computer programs written for the projects, were also published in my thesis.

In GIS, there are two commonly used methods, that is, digitizing and scanning, to convert paper maps into digital maps and their associated datasets (also known as data layers, or themes). A digitizer is a piece of electronic hardware, consisting of a digitizing table (also called a tablet or a board) and a cursor (also called a mouse or magnetic pen). A digitizing table has a wire grid underlying its surface, serving as the table's coordinate system. A paper map is mounted onto the digitizing table and a GIS operator traces the spatial features on the paper with a cursor. Before digitizing, a map coordinate system or projection has to be defined and registered with control points on the map. These control points are usually the corners of the map if the whole map is going to be digitized, or the corners of the area of interest inside the map. There are two commonly used digitizing modes: point mode and stream mode. When tracing the features on the map in a point mode digitizing process, each click on the feature with the cursor is treated as a point (also called a vertex) by GIS, and its coordinates (in either easting and northing, or longitude and latitude format) are recorded. In stream digitizing mode, points are recorded at predefined regular intervals of time or distance continuously while tracing the map features without manually clicking with the cursor. This technique saves the time and work of manual clicking. There are a variety of other techniques to assist the GIS operator in the digitizing process, which is very labor-intensive and attention demanding. While digitizing, GIS allows the operator to see the digitized features on a computer screen, correct errors, and/or edit the features. Some advantages of manual digitizing are as follows.

- Any map—regardless of its contents and quality—can be digitized manually, while not all maps can be scanned, especially those maps with complex spatial features, symbols, labels, handwriting, fading ink, folding lines, worn and torn parts, large sizes, and so on.
- Both vector paper maps and raster images can be digitized.
- In general, digitizing devices are affordable, reliable, and accurate. Therefore, less capital investment is required and data layers produced with digitizing are of high quality and accuracy.
- It is relatively easy to train less expensive workers to perform digitizing tasks to save labor costs because digitizing is labor intensive. Workers can quickly learn digitizing tools and techniques to improve work efficiency and reduce data input errors.
- An operator can see and correct errors on computer screen while digitizing.
- For data input with lower accuracy requirements, features on a paper map can be estimated and digitized directly on a computer screen instead of using a digitizing table. This method is commonly called head-up digitizing, while digitizing on a table is called head-down digitizing. Head-up digitizing is usually used in tracing geographic features from a map with aerial, satellite, or digital raster graphics (DRG) imagery background, with the computer screen also showing the similar imagery. The head-down digitizing method is commonly utilized for complex vector maps.
- In general, less post-processing time and effort are needed to identify and correct data entry errors, edit features, and build topology.

With a scanning method, paper maps can be scanned into digital data quickly without much human involvement. This method could be an efficient alternative for maps in good quality and condition, with relatively easy-to-recognize spatial features, clean and visible lines, fewer symbols and labels, no paper folding lines, and no worn or torn parts or other defects. However, at the technology's current level, scanning devices are not as smart as human eyes in recognizing spatial features (e.g., points, lines, and polygons) and differentiating them from symbols, labels, and map defects. Therefore, some amount of data-capturing errors will inevitably be introduced during the scanning process, depending on the contents and qualities of the hard copy maps. Although it is faster to scan a map than digitize it, it usually takes longer to post-process scanned data than digitized data. For a map with complicated features, symbols, and labels too many errors could be generated during the scanning process and it would take too much time to edit the data and build a correct topology than simply to re-digitize it. On the other hand, for a map with not many features, it is more efficient digitizing it than scanning it and editing the data errors. For a map in poor condition, some features can be lost during the scanning process. They have to be manually added during the post-data-processing process. Due to the technical limitations of scanning and also the fact that large size scanners are generally more expensive than digitizing devices, the digitizing method is more commonly used than the scanning method in converting hard copy maps into digital GIS data and maps. However, some special spatial data companies and government agencies use scanning technology to process large volumes of hard copy maps.

The selection of the better data-converting method then depends on the quality and size of the paper map, the complexity of the features, the symbols and labels on the map, and the availability and quality of the hardware, such as the size and accuracy of a digitizing tablet or a scanner. In general, if the quality of a paper map is reasonably good, without wrinkles and/or other defects, and containing relatively simple features (i.e., not too crowded), it is more efficient to scan the map and convert it into vector data layers. Although scanning is quicker to begin with, it may take a longer time to clean up and edit the vectors converted from the scanning process because a certain amount of errors are always introduced during a raster (i.e., image) to vector conversion process. The more complicated the paper map is, the more errors are introduced into the converted vector data layers. If errors are not corrected completely in the vectors the topology cannot be built correctly, which will result in some features possibly being missing. Therefore, the GIS data layers generated from the scanning process are not correct or accurate. In contrast to the scanning method, digitizing can be used for any hard copy maps, as long as the map fits on an available digitizing tablet. If it is digitized carefully, there is less cleaning and editing work required on the vectors and thus less effort expended in building a correct topology, too. However, it is time-consuming and tiring to digitize a map with a lot of features. Choosing the best method can save lots of time and money in converting a paper map into a digital map and its associated GIS data layers.

Since the quality of the hand-drawn 1970s Mylar maps was not good, and the maps are full of geologic features, symbols, and labels, I decided to use a digitizing method instead of scanning the map. Also there was no scanner on campus at that time which was large enough to scan the big Mylar maps. Further, I chose to divide the large 1:50,000 Mylar map into four 7.5 minute quadrangles following the boundaries of the four USGS topographic quadrangles as the map was originally composed. Otherwise, it would take too long to digitize the huge map with its many features. Also, it would be too complicated to clean the features of the whole map and build the topology correctly, especially for an inexperienced GIS learner.

To check that my choice of digitizing method was a better approach, I scanned a small portion of a Mylar map as an experiment. As anticipated, the vector features converted from the scanning process were full of errors. It took longer to clean up and edit the features than to re-digitize them. So, my first step of choosing digitizing was proven to be correct. It was a good step in the right direction.

However, being one of the early pioneers in this interdisciplinary research and experimentation with GIS technology in geosciences, I soon encountered many challenges and problems. These challenges included choosing an appropriate coordinate system for the maps and data layers; dealing with special geologic symbols on the paper maps; finding a more efficient digitizing procedure; matching the edges of the four 7.5 minute quadrangle maps and joining them together; building the topology of the data layers; and ensuring data quality.

In the early and mid-1990s, GIS was still a relatively new technical field that was not widely used and adopted in other academic disciplines, including Earth sciences. Although there were experienced GIS professionals and geology experts on campus, there was no expert who knew both geology and GIS technologies well enough to guide me through the whole process. There was not much relevant literature on both GIS and Earth sciences to refer to either. I had to learn GIS theories and techniques, understand the geologic data and maps, and perform various experiments to overcome the challenges and solve the problems one by one.

In one year or so, I successfully converted the five 1:24,000 scale and one 1:50,000 scale geologic maps, hand-drawn on Mylar sheets, into digital maps and their associated GIS data layers, containing information about the geologic features, symbols, and labels. When I presented these colorful digital geologic maps, as shown in Figure 4.2, to the 10th Annual Graduate Student Seminar of the Department of Geological and Atmospheric Sciences, Iowa State University on March 28, 1998, the professors and students were impressed by how easy it was to read and understand the digital GIS maps with their visually appealing colors, symbols, and labels, especially when compared with the black and white hard copy maps with their poor line quality and crowded symbols and labels. They were further amazed when I demonstrated to them how the digital maps and data layers, combined with other related data layers (such as aerial/satellite images, elevation data, and regional geologic structures) could be manipulated and analyzed to better understand geologic units and structures and their patterns, trends, and relationships on local and regional scales by turning data layers on and off, querying certain features out and highlighting them, changing colors and symbol sizes based on their values, and plotting geologic measurement

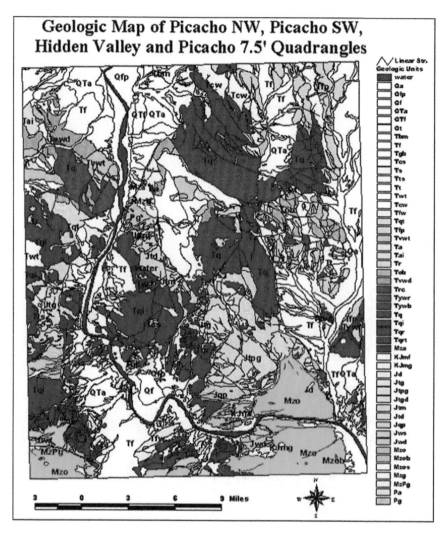

FIGURE 4.2

A digital GIS map converted from the hand-drawn 1:50,000 "Geologic Map of the Picacho, Picacho NW, Picacho SW, and Hidden Valley 7.5′ Quadrangles, Arizona and California" by D. R. Sherrod, R. M. Tosdal and G. B. Haxel. The long thin blue polygon running from north (top) to south (bottom) and then toward east (right) represents a small river. Other colored polygons are geologic units, labeled with their abbreviated names. This digital geologic map was created with the ESRI's ARC/INFO workstation version 7 GIS software, running on a Unix operating system, in late 1997.

data as symbols on the maps automatically in real time, as illustrated in Figures 4.2 through 4.4.

Figure 4.2 is a GIS map showing the geologic units of Picacho NW, Picacho SW, Hidden Valley and Picacho 7.5 minute quadrangles in southwestern Arizona and southern California. This digital geologic map was generated with the GIS data layers digitized from the 1:50,000 scale hard copy map

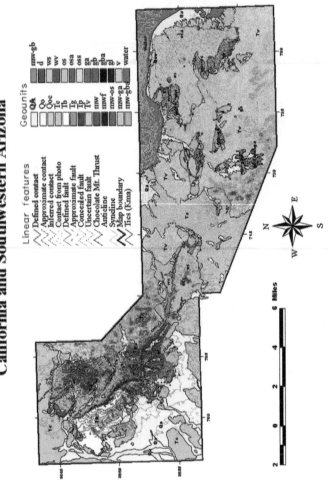

FIGURE 4.3

A digital GIS map converted from the five 7.5 minute (1:24,000) quadrangle hard copy geologic maps of southern California and southwestern Arizona by D. R. Sherrod, R. M. Tosdal, G. B. Haxel, C. Jacobson and others. The colored polygons are geologic units labeled with their abbreviated names. It also contains various types of faults, anticline and syncline axis lines and topographic contours generated from USGS DEM datasets. This digital geologic map was created with the ESRI's ARC/INFO workstation version 7 GIS software, running in Unix operating system, in late 1997.

FIGURE 4.4
Hidden Valley 7.5' quadrangle 3D geologic map, southwestern Arizona. The 3D topographic surface model was generated from the USGS DEM (Digital Elevation Model) data. The hard copy map was originally hand-drawn by D. R. Sherrod, R. M. Tosdal and G. B. Haxel. The long thin white polygon running from west (left) toward east (right) on the lower portion of the map is a small river bed. Other colored polygons are geologic units. This digital GIS 3D geology map was created with ESRI's 3D Analyst in early 1998.

entitled "Geologic Map of the Picacho, Picacho NW, Picacho SW and Hidden Valley 7.5' Quadrangles, Arizona and California," which was hand-drawn by D. R. Sherrod, R. M. Tosdal, and G. B. Haxel in the 1970s, as shown in Figure 4.1. This digital geologic map and its data layers can be manipulated and analyzed to reveal patterns and relationships and to interpret geologic structures. It is much easier to visualize and understand a digital map than its original hard copy version, especially for students.

Figure 4.3 is a large digital geologic map created by digitizing and joining five 7.5 minute, 1:2,400 scale quadrangle maps to display the regional geology of the southern California and southwestern Arizona areas. The original five 1:24,000 scale hard copy geologic maps were hand-drawn by the USGS geologists Gordon Haxel, David Sherrod, and Richard Tosdal in the 1970s.

The large geologic map consists not only of geologic units but of geologic structures such as faults, axis lines of anticline and syncline, and elevation contours that were analyzed from USGS Digital Elevation Model (DEM) datasets. Since this geologic map contains so many data layers/features, it is hard to read the detailed information on this small map. However, when it is printed on a large sheet of paper, the detailed data will show up clearly. By manipulating and analyzing these data layers and features, it is much easier for students and geologists to see the big picture of the regional geology and structures and discover patterns, relationships and trends that would be hard to achieve with hard copy paper maps.

Figure 4.4 is a three-dimensional (3D) geologic map generated by overlaying a geologic units data layer onto a 3D topographic model created from USGS DEM datasets. With this 3D geology map it is much easier to visualize the geologic features and structures of the study area, both locally and regionally, compared with flat 2D geologic maps.

Although my professor and the geologists who originally hand-drew the geologic maps reviewed the digital GIS maps carefully and confirmed to me that they were correct, accurate, and much better than the paper maps, I was still not certain that I had performed every step accurately in the whole digitizing and data post-processing process. This was because I could not find anyone on campus to evaluate my work process and the GIS products from both geology and GIS perspectives. I did extensive research on GIS data quality assessment and quality assurance and quality control (QA/QC). Referring to widely accepted data QA/QC theories, standards and procedures, I reviewed all the GIS datasets and maps, corrected the errors identified, and further improved the data layers and maps. On April 25, 1998, I presented a paper to the 110th Annual Meeting of the Iowa Academy of Science, in Mason City, Iowa, entitled "Applications of GIS in Structural Geology: A Case Study on the Structural Geology of Southern California and Southwestern Arizona," based on these GIS maps and the datasets. It was a great success with positive feedback and follow-up discussions.

Since my graduation in late 1998, Professor Jacobson and his students have further improved and updated the GIS maps and datasets with their new research results and data. In 2010, they presented the 1:50,000 GIS map to the prestigious Geological Society of America (GSA) Annual Meeting, which ran from October 31 to November 3, 2010, in Denver, Colorado, in a session entitled "Digital Geologic Map of the Picacho, Picacho NW, Picacho SW and Hidden Valley 7.5' Quadrangles, Arizona and California."

In order to allow more interested geology professionals and students to use the GIS maps and data layers, especially those users without GIS knowledge and any commercial GIS software license, I also wrote a free GIS viewer software application. The free GIS application was named "Geologic Map Viewer". It contains some basic tools that allow users to add data layers, zoom in/zoom out on the map, pan the map around, turn data layers on and off, change colors and symbols, label features, and perform simple edits to the data layers. It was written with Visual Basic 5.0 and ESRI MapObjects 1.2 for desktop computers. The source code of the free GIS viewer application was published in my thesis and distributed over the Internet. Figure 4.5 is

FIGURE 4.5

A screenshot of the free GIS application named Geologic Map Viewer, written in 1997 with Visual Basic v. 5.0 and ESRI MapObjects v. 1.2 for desktop computers. The long thin yellow polygon running roughly from north (top) toward south (bottom) on the west portion of the map represents a small river. Other colored polygons are geologic units. This Geologic Map Viewer software allows users to add data layers, zoom in or zoom out on the map, pan the map around, turn data layers on and off, change colors and symbols, label features, and perform some simple edits to the data layers.

a computer screenshot of this free GIS map and data viewer application. It shows geologic units, faults, and other features.

Case Study 4.2: Creating 3D Maps and Studying Them with Virtual Reality (VR) Technology

As a class project in my computer graphics and virtual reality course, I experimented with utilizing virtual reality (VR) to perform 3D modeling and visualization of geologic strata and structures. The GIS maps and datasets of the study areas in southern California and southwestern Arizona were used in the VR experiments.

Virtual reality is a computer-simulated 3D environment. Immersed inside a VR environment, users feel as if they are in a real-world environment and are able to manipulate simulated objects and perform actions interactively with the VR computer system using special equipment such as displaying and viewing devices (e.g., head-mounted display, booms, and large projection screens), tracking and interactive input devices (e.g., 3D mouse and sensing gloves), and sound devices. VR technology is widely used a variety of fields, for example, terrain visualization for flight and combat training purposes, medical education, outer space exploration, architecture, natural resource exploration, scientific research and education, and entertainment. However, a VR system is very expensive to set up and operate. Not many organizations can afford this high-end technology or have the expertise to operate it.

Due to the great efforts of Dr. Carolina Cruz and other faculty, a CAVE Automatic Virtual Environment (CAVE) VR system was installed on the Iowa State University campus. The projection-based CAVE hardware system is composed mainly of four large screens that give the user a highly immersive feeling, stereo glasses, wands (3D mouse), sensing gloves, tracking systems, an audio system, and high-end computer workstations. The operation of the system is controlled by the CAVE library of C functions and macros. The CAVE VR system offered a great opportunity for me to experiment with 3D visualization of the geologic features in the southern California and southwestern Arizona region.

I built a 3D terrain model of the Picacho SW 7.5 minute quadrangle in southern California from the U.S. Geological Survey Digital Elevation Model (DEM) data with OpenGL programming language and imported it into the CAVE. Figure 4.6 is a picture of the 3D terrain model and simulated virtual walking inside the 3D terrain model in the CAVE.

By manipulating the simulated immersive display, head-mounted display, sensing gloves, and wand, the user was able to navigate inside the 3D terrain model in the VR environment, feeling as if he was walking the real

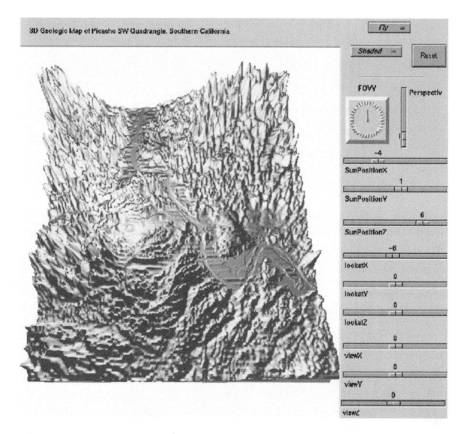

FIGURE 4.6
A 3D terrain model of the Picacho SW quadrangle in southern California. It was generated from USGS DEM data with OpenGL programming language and imported it into the CAVE VR environment for simulated walking and flying experiments. The small river bed running roughly from north (top) to south (bottom) and then toward east (right) is more visible on this 3D terrain model.

mountains in that region. The user could also fly over or inside the 3D terrain model, as illustrated in Figure 4.7.

Since the file size of the 3D terrain model was very large it consumed too much computing resources, a multiresolution technique was implemented to reduce lagging effects during the flying simulations. When the user was far away from the model, the resolution of the overall 3D terrain model was reduced to increase the speed of the rendering and simulating process so that the simulated flying was smoother and felt more realistic. However, when the user flew close to the 3D terrain model the resolution of the local area was increased so that the user could see the details of the terrain without noticing the low resolution in other areas not seen, similar to a real-world experience.

FIGURE 4.7
A simulated flight over the mountains along a valley inside the 3D terrain model of the Picacho SW quadrangle in southern California, in the CAVE. The red ball with a black dot is the simulated head (with a black eye) of the flying person. The valley on the center of this the 3D terrain model is the river bed as shown on other maps.

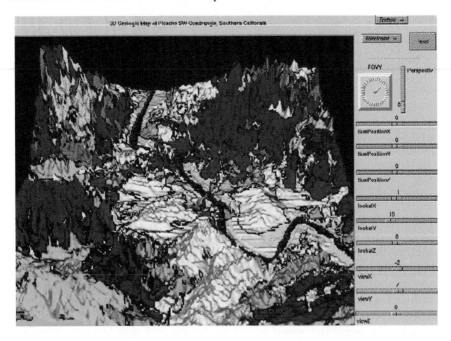

FIGURE 4.8
A 3D geology model of the Picacho SW quadrangle in southern California. It was generated by texture mapping a geology map onto the 3D terrain model, and imported it into the CAVE Virtual Reality Environment for simulated walking and flying experiments. The long thin blue polygon running from north (top) to south (bottom) and then toward east (right) represents a small river. Other colored polygons are geologic units.

When the walking and flying 3D terrain model experiments in the CAVE proved successful, I texture-mapped the geology units and structures GIS data layers onto the 3D terrain model, making it a 3D geology model. I then performed similar walking and flying experiments. By manipulating lighting, shading, and texture mapping techniques, it felt like walking or flying inside the real mountains and valleys covered with real geologic units and structures.

Figure 4.8 is a picture of the 3D geology map in the CAVE for simulated walking and flying. It was a great way to visualize geologic structures in a simulated real-world environment. The source code of that VR project and data was published in my master's thesis.

Case Study 4.3: Processing and Mapping Geologic Data More Efficiently with GIS

During the extensive research and study of the structural geology of southern California and southwestern Arizona over the past years, a huge amount of field data, especially measurements of geologic structures, have been collected mainly by Professor Carl Jacobson, his students and others. These measurements data are very important in understanding current geologic structures and interpreting the history and tectonic mechanisms of the local study areas and the western region of the United States too.

Structural geology is a branch of Earth science that studies the form, arrangement, and internal structure of rocks in 3D spaces, uncovers the history and mechanisms of rock deformations, and predicts possible future movements. By measuring rock geometries, geologists try to understand and construct current geologic structures and interpret their deformational histories on local and regional scales.

Strike and dip measurements are commonly used to define the orientation and inclination degrees of a planar structure in geology. The strike line of a planar feature is a line representing the intersection of that feature with an imaginary horizontal plane. A strike measurement can be expressed either as a quadrant compass bearing of the strike line, for example, S45°E, or as a number, ranging from 0° to 360°, clockwise from true north, for example, 135°. The dip line is at a right angle to the strike line. It represents the magnitude of the inclination of a planar feature below an imaginary horizontal plane, ranging from 0 to 90, as illustrated in Figure 4.9.

In structural geology, lineations are linear structural features or elements within rocks formed by tectonic, mineralogical, sedimentary, or geomorphic processes. There are numerous types of lineations, such as intersection lineations formed by the intersection of any two surfaces, crenulation lineations formed by the intersection between fold hinges and foliation, mineral

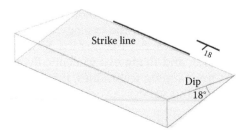

FIGURE 4.9

A diagram illustrating the strike and dip of a planar geologic feature, such as the surface of a geologic unit, fault, dike, etc. The longer line represents the strike of the planar feature and the shorter line represents the dip 18° below an imaginary horizontal plane in this example.

LOCATION	TRND	PLNG	X	Y
1	299	23	711.985	3655.722
2	331	35	711.985	3655.722
3	263	41	711.985	3655.722
4	279	16	711.987	3655.789
5	280	36	711.926	3655.830
6	278	38	711.919	3655.904
7	270	16	711.904	3656.036
8	306	27	712.090	3655.485
9	308	13	712.090	3655.485
10	306	35	712.090	3655.485
11	312	17	712.165	3655.508
12	280	43	712.165	3655.508
13	261	26	712.165	3655.508
14	309	52	712.165	3655.508
15	314	21	712.165	3655.508
16	264	21	712.526	3655.427
17	233	9	712.459	3655.472

FIGURE 4.10

A small portion of a lineation measurements data table. Column "TRND" contains tread measurements in 0°–360° range. Column "PLNG" contains plunge measurements in 0°–90° range. X and Y columns contain coordinates of the locations of the measurements.

lineations formed by the preferred alignment of minerals due to deformation or recrystallization during deformation, and stretching lineations formed by the elongation of minerals due to stretching deformation. The spatial orientation of a lineation is defined by its bearing and plunge. The plunge is the inclination of the lineation below a horizontal plane (ranging from 0° to 90°). The bearing, also known as the trend or plunge direction, is the direction of the lineation in the direction of the plunge (ranging from 0° to 360°).

Figure 4.10 is a small portion of a lineation measurements data table queried out from a large geology database. It contains the coordinates of the locations of the measurements in the X and Y columns (also known as fields), the bearing measurements (also known as trend) ranging from 0° to 360° in the TRND column, and the plunge measurements of lineations ranging

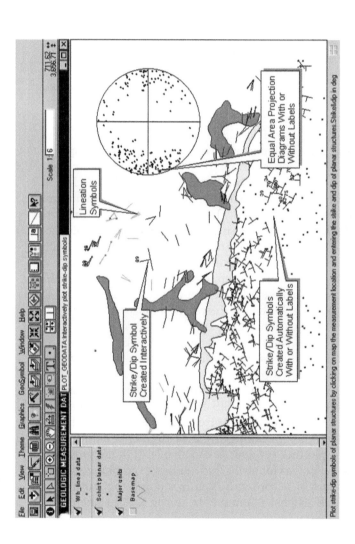

FIGURE 4.11

A screenshot showing how the "PLOT_GEODATA" GIS extension was used to plot strike/dip and lineation symbols onto a geology map either automatically or interactively. When the extension is loaded, three command buttons and two tool buttons will appear on the top of the window. When a mouse pointer is on a command or tool button, a brief description of it shows up in a yellow banner next to the button, e.g., the "PLOT_GEODATA: Interactively plot strike-dip symbols" banner in the UGI on the top. Also, detailed instructions on how to use the button appear in the bottom of the window, e.g. "Plot strike-dip symbols of planar structures by clicking on the map the measurement location and enter the strike and dip of planar structures..." The button (with a strike-dip symbol and "a" on it) can be used to plot strike/dip symbols automatically by reading their measurements from a data table. The strike/dip button (without "a" on it) can be used to plot strike/dip symbols interactively. The button with a line on it can be used to plot lineation symbols automatically. The button (with the stereonet diagram on it) can be used to automatically plot lineations in a Schmidt equal area projection diagram by reading trend and plunge data from a data table. The points on the map are strike/dip and lineation measurement locations. The colored polygons are geologic units. Other geologic units and features were turned off.

from 0° to 90° in the PLNG column. The format of a strike and dip measurements data table is similar to this lineation measurements data table, with the STRIKE and DIP data columns replacing the TRND and PLNG columns, respectively.

By displaying strike/dip and lineation symbols and their measurements on a geology map it is much easier for geologists to visualize their patterns and relationships and interpret regional geologic structures and their histories. Due to the huge amount of measurements data collected over so many years from the large study area, it would be very time-consuming and error-prone to plot the symbols manually onto geologic maps one symbol at a time. After extensive research and experiment, I wrote a GIS extension named "PLOT_GEODATA" to plot strike/dip and lineation symbols on geologic maps automatically by reading their measurements from data tables or databases, as illustrated in Figure 4.10. This GIS extension was compiled from multiple individual programs written to automate certain geologic mapping processes. When this GIS extension is installed, three command buttons and two tool buttons will appear in the graphical user interface (GUI) at the top of the window. The buttons can be used to plot strike/dip and lineation symbols onto geologic maps either automatically by reading their measurements from data tables or interactively with the user clicking measurement locations on the map and entering their measurement values one by one. With large amounts of measurements the automatic plotting button definitely is the best option to save time and prevent human data entry errors. The interactive plotting button is useful when only a few new measurements need to be added to the map to update it. When the interactive plotting tool button is clicked it will prompt the user to either click on the map or key in the coordinates for the chosen location. Then it asks the user to key in the location's measurement values in the order prompted, for example, strike measurement value (in 0°–360° range) and dip angle (in 0°–90° range) for a strike/dip symbol, or trend measurement value (in 0°–360° range) and plunge angle (in 0°–90° range) for a lineation symbol. This GIS extension can also be used to automatically plot lineation measurements of trend and plunge in a Schmidt equal area projection diagram by reading the measurements data from a table or database. Besides plotting symbols, it can also label them with their measurements, such as dip angles for strike/dip symbols and plunge angles for lineation symbols. Figure 4.11 shows the command and tool buttons of this "PLOT_GEODATA" GIS extension and the strike/dip and lineation symbols plotted on a geology map.

5

Data Modeling and Visualization

Some environmental problems can be complicated, such as large-scale explosions in chemical facilities; oil spills/gas leaking from wells, tankers, storage, and distribution piping networks; abandoned conventional or chemical munitions from past military trainings or wars; damaged landfills; defective sewer or storm water systems; improper disposal of hazardous wastes; smog; greenhouse gases; and so on. They may acutely or chronically impact human health, the atmosphere, soil, surface water, groundwater, and ecosystems. A sufficient amount of data/information is required to be collected, processed, and analyzed to understand the problems accurately and find the most appropriate and effective mitigation and remedial strategies. Sound decisions are made based on an accurate understanding of large-quantity and good quality data.

In the past 20 years or so, due to the significant improvements in data-capturing technologies, such as remote sensing with satellites and airplanes, global positioning systems (GPSs), Geographic Information Systems (GISs), and others, data collection has become more automated and efficient. Compared with the manual data-collecting era—which was not too long ago—nowadays there is no shortage of data, and in some cases there is maybe too much data to sort out and understand them correctly. Although data-capture technologies have advanced dramatically in recent years, the human intellectual capability of visualizing, understanding, and interpreting large quantities of complicated datasets does not keep up with the swift advances of the data-collecting technologies. Therefore, collecting a large amount of data does not automatically mean achieving better understanding of the problem, making educated decisions, and finding the best solutions. Instead, dealing with too much data can be intimidating, confusing, and possibly misleading. Important patterns, relationships, trends, and other useful information can be hidden behind the boring, overwhelming, and complicated numbers. Previous chapters discuss how GIS techniques can be utilized to process and analyze large quantities of data to help scientists and decision-makers understand data, and make sound and educated decisions. This chapter will explore how to use GIS and other related technologies to model and visualize large environmental and geological data to uncover hidden patterns, trends, relationships, and other information.

Data modeling is a mathematical and statistical process that sorts through complex data to find the relationships, patterns, and trends and translates them into easy-to-understand information formats such as charts, two-dimensional or three-dimensional graphics (also known as models), and summary tables. Data visualization, closely related to data analysis and modeling, is both a science and an art, which sieves out, sorts, manipulates, and displays complicated source data in condensed, simplified, and therefore more easily understandable formats such as graphics, charts, pictures, illustrations, and animations. Its primary goal is to help users see through massive and complex data; uncover patterns, trends, and relationships; and tell stories. It converts complex data into more accessible, understandable, usable, and meaningful information. However, caution needs to be taken to ensure that the results of modeling and visualization are true and realistic representations of the source data. Otherwise, if the source data are overly simplified, distorted, and/or misrepresented, the results could be misleading or even wrong, no matter how visually appealing they are.

Although GIS offers powerful modeling and visualization capabilities, there are many other data modeling and visualization software applications designed for various specific purposes, such as EquIS, EnviroInsite, Groundwater Modeling System (GMS), Environmental Visualization System (EVS), RockWorks, and many more. They are widely used in modeling and visualizing geology, hydrology, contaminated soil, and groundwater plumes. GIS can be integrated with a variety of data modeling and visualization software applications through extensions, plugins, or custom applications so that modeling and visualization can be executed inside a GIS application seamlessly. GIS can also allow other data modeling and visualization applications to access data stored in GIS databases directly instead of preparing separate datasets. Some of the commonly used data modeling and visualization software applications are briefly discussed in the following.

EquIS

Environmental Quality Information System (EquIS) is an environmental database management software package developed by EarthSoft. The desktop version of EQuIS uses the Microsoft Access database engine and the Microsoft VB.NET programming environment, while the high-end EQuIS Enterprise is based on the SQL Server or Oracle data platform, which allows regional or nationwide access to data over the Internet.

EquIS streamlines environmental data management for biology, chemistry, geology, geotechnical, water, limnology, air quality, and other data types and contains various data analysis, mapping, modeling, and visualization functionalities and tools. Due to its open systems (also referred to as open platforms) business model, EquIS allows and attracts third parties (including commercial developers and ordinary users) to make software products that plug into or interoperate with EquIS to further enhance, expand, and/or customize its capabilities. A good example is the development of GIS tools and extensions to integrate EquIS with ESRI GIS software products (from its old-version ArcView 3.x to current ArcGIS 10.x) to combine ESRI's powerful mapping and data analysis capabilities with EquIS's advanced environmental data management functionalities using the same stored data, eliminating the complicated and time-consuming process of exporting, converting, editing, reformatting, and importing data. Besides tools and extensions for GIS, there are a variety of other third-party interfaces developed to enable EquIS to interoperate with other leading groundwater, subsurface, and statistical software applications for extended data analysis, modeling, and visualization. Following is a list of the major software applications that EQuIS can interoperate with:

- *ArcGIS*: EQuIS for ArcGIS is an extension for the ESRI's ArcGIS family desktop applications, including ArcView, ArcEditor, and ArcInfo. With the EQuIS for ArcGIS extension, GIS users can query data from EQuIS databases and use them in the GIS environment for data analysis, modeling, and mapping.

- *EnviroInsite*: Developed by HydroAnalysis, EnviroInsite is a groundwater data modeling and visualization software package, capable of analyzing geochemistry using piper diagrams, Stiff diagrams, and Schoeller diagrams; visualizing project site geologic and hydrogeologic settings using fence diagrams, strip logs, and geologic modeling; and revealing complex spatial and temporal relationships in environmental data using pie charts, 3D surfaces, and particle tracking.

- *Environmental Visualization System (EVS Pro)*: Developed by C Tech, EVS is a software package for data analysis, visualization, and animation.

- *gINT*: Developed by Bentley, gINT is a geotechnical and geo-environmental software product, especially useful in subsurface reporting for soils, borelogs, lab tests, and other data for environmental, geophysical, petroleum, and mining projects.

- *Groundwater Modeling System (GMS)*: Initially developed by Brigham Young University and funded primarily by The U.S. Army Corps of Engineers, GMS is a software package used to create groundwater and subsurface simulations in a 3D environment. Since April 2007, GMS development has been taken on by Aquaveo.

- *Groundwater Vistas (GV)*: Developed by Environmental Simulations Inc. (ESI), Groundwater Vistas is a software package for 3D groundwater flow and contaminant transport modeling. It is a pre- and post-processor for MODFLOW models.

- *RockWorks*: Developed by RockWare, RockWorks is a software program widely used in the petroleum, environmental, geotechnical, and mining industries for subsurface data analysis, modeling, and visualization. It contains tools such as maps, logs, cross sections, fence diagrams, solid models, and volumetrics.

The third-party interfaces make EquIS a popular environmental data management software application widely used by numerous government agencies (such as U.S. EPA, U.S. Army Corps of Engineers, some U.S. states, Kuwait EPA, and others), multinational corporations and industrial property owners, consultants, and analytical laboratories around the world. Following are a few useful features of EquIS:

- Better management of environmental liabilities
- Improves data integrity and reduces costs
- Works with Oracle, Sql Server and gi
- Supports workflow automation via Outlook, Excel, and other tools
- Improves retention of existing data
- Improves project transitions for new internal and external staff
- Lowers costs and reduces the hours needed to input, create, and distribute information
- Promotes high standards and consistency in data collection and reporting
- Monitors project progress and data from a central location
- Increases your ability to share documents and data through web portals
- Easy communication with project teams
- Improves project decisions with management review processes
- Makes transitioning between contractors easier
- Eliminates redundant legacy systems.

Groundwater Modeling System (GMS)

GMS was originally developed in the late 1980s by the Engineering Computer Graphics Laboratory of Brigham Young University, with funding coming mainly from the U.S. Army Corps of Engineers (USACE). The Engineering Computer Graphics Laboratory was renamed the Environmental Modeling Research Laboratory (EMRL) in September 1988. Later, in 2007, the software development team at EMRL formed a private software development company called Aquaveo to continue to maintain and develop GMS and other software products. Since the development of GMS was funded by the U.S. Army

Corps of Engineers, an agency inside the U.S. Department of Defense (U.S. DoD), it is also called the Department of Defense Groundwater Modeling System (GMS). The U.S. Army, in partnership with the U.S. Environmental Protection Agency, the U.S. Nuclear Regulatory Commission, and various academic partners, has developed the DoD GMS. Therefore, GMS is free for use by the U.S. DoD and other related government agencies as well as authorized DoD contractors working on DoD projects or programs.

GMS is an intuitive and comprehensive software package with a variety of modules used to construct groundwater and subsurface geologic models and simulations in a 3D environment, such as 3D stratigraphy models and 3D contaminated groundwater plumes as well as groundwater flow, fate, and transport modeling. GMS's GIS tools and its conceptual model approach make it easier and faster for GIS users to build groundwater flow and contaminant fate/transport models in complex hydrogeological settings, transfer lithology and hydrogeology data into the models, convert data from a GIS database into model boundaries, adjust boundary conditions, assess modeling errors, and evaluate the efficiency of remediation systems. GMS contains a variety of subsurface modeling tools, such as creating iso-surfaces to visualize contaminated groundwater plumes in 3D spaces, constructing 3D stratigraphy models and cutting cross sections/fence diagrams anywhere through them, and performing 2D or 3D geostatistic analysis using the Kriging, IDW, and Natural Neighbor methods. In addition, GMS 3D has various data visualization capabilities, such as enabling photo-realistic renderings; draping images over the 3D model and adjusting their opacity; interacting with models in a true 3D environment; generating animations for PowerPoint and web presentations; and adding north arrows, scale bars, reference images and company logos. The current version of GMS v.10.0 provides an interface for a variety of models. Following are some of the supported models:

- ADaptive Hydraulics Modeling system (ADH): ADH is a 2D flow and sediment transport modeling application, used in riverine modeling.

- FEMWATER: FEMWATER is a 3D finite element groundwater and contaminant transport modeling computer program, used to simulate density-driven coupled flow and contaminant transport in variably saturated media and zones.

- MODAEM: MODAEM is an analytic element modeling application, used in groundwater flow and transport modeling. Different from finite element models, analytic element models are only defined by boundary conditions, source and sink terms, and material property zones which are represented by points, polylines, and polygons.

- MODFLOW: Originally developed by the U.S. Geological Survey (USGS) in the early 1980s using the Fortran programming language, MODFLOW is a 3D modular finite-difference groundwater flow model. It is a free open-source software program. Since then, several MODFLOW versions (e.g., MODFLOW-88, -96, -2000, -2005, -LGR, NWT, -USG, -OWHM, GSFLOW, GWM, CFP and SWR) have been developed and released, and many commercial and noncommercial graphical user interfaces have been developed too. It is so widely adopted and used that MODFLOW is now considered an international standard for simulating and predicting groundwater conditions and groundwater/surface water interactions. Due to its open-source and modular structure, MODFLOW serves as a robust framework for integration of additional simulation capabilities. Programs from the MODFLOW family incorporate a wide range of modeling and simulating capabilities, such as simulating coupled groundwater/surface water systems, solute transport, variable-density flow, aquifer system compaction and land subsidence, parameter estimation, groundwater management, and many other capabilities. Its current version is MODFLOW-2005 v.1.11.00. The current version GMS v.10.0 supports various structured and unstructured MODFLOW grids to suit for different modeling needs.

- MODPATH: MODPATH is a particle tracking post-processing software package used to calculate 3D flow paths using output results from steady state or transient groundwater flow simulations by MODFLOW and/or other datasets. It was developed by the U.S. Geological Survey using the Fortran programming language. Particle tracking can be performed either forward or backward in time sequences. Particle paths are computed by tracking the possible movement of imaginary particles in a groundwater system from one grid cell to another adjacent cell until the particles reach a boundary, an internal sink/source, or satisfy some other termination conditions. MODPATH is useful in delineating capture zones, source of contaminants, areas of influence for wells, and so on.

- MT3DMS: MT3DMS is a modular 3D multispecies transport model for simulating advection, dispersion, and chemical reactions of contaminants in groundwater systems, using the heads and cell-by-cell flux terms computed by MODFLOW during the flow simulation. It is widely used in contaminant transport modeling and remediation assessment studies. Its earlier MT3D version was developed by Chunmiao Zheng in 1990. Since the development was partially funded by the U.S. Environmental Protection Agency (USEPA), MT3D was given a public domain release. MT3DMS is an improved version of MT3D, with some new features and enhanced functionalities. The development of MT3DMS was funded by the U.S. Army Corps of Engineers Waterways Experiment Station under the Strategic Environmental Research and Development Program (SERDP).

- RT3D: RT3D is a Fortran-based software package for simulating 3D, multispecies, reactive transport of chemical compounds in groundwater, developed by Pacific Northwest National Laboratory (PNNL), an agency of the U.S. Department of Energy (DOE). It is a modified version of MT3D, with expanded reaction simulating and modeling capabilities. RT3D is useful in simulating natural attenuation and accelerated bioremediation as well as quantitatively evaluating contaminant precipitation, absorption, and migration in subsurface media.

The Groundwater Modeling System supports a wide range of coordinate systems/projections and a variety of data file formats, including GIS vector and raster data files, CADD files, and output data files of MODFLOW and many other models. Following is a list of the data formats supported by the Groundwater Modeling System:

- Topographical maps and elevation data, such as Digital Elevation Model (DEM), National Elevation Dataset (NED), Digital Terrain Model (DTM), and so on
- Borehole data, including stratigraphy and geophysical data
- Native MODFLOW files
- MODFLOW files from Visual MODFLOW, Groundwater Vistas, and others
- Web data services, such as TerraServer
- ArcGIS geodatabases, shapefiles, and numerous raster data files
- CAD files, including AutoDesk's AutoCAD (.dwg, dxf) and Bentley's MicroStation (.dgn) formats
- Delimited text files and spreadsheets
- Worldwide projection support, including commonly used Cartesian; (i.e., projected) and Geographic Coordination Systems

Environmental Visualization System (EVS Pro)

The Environmental Visualization System (EVS Pro) was originally developed in 1991 by C Tech for advanced gridding, geostatistical analysis, 3D modeling, and visualization. In addition to GIS capabilities, EVS Pro has a variety of other functionalities and features, such as volume rendering, generating animation, terrain flyover, geologic modeling, interactive fence diagrams, 4DIM/VRML (used for creating 3D PDFs and 3D Printing) output, database connectivity, aerial photo texture mapping, handling complex geologic layers, and others. It is especially useful in modeling and visualizing groundwater flow and contamination conditions in groundwater and subsurface soil. From December 2015 onward, C Tech terminated both the Environmental Visualization System (EVS Pro) and the Mining Visualization System (MVS) software products and replaced them with the newly developed Earth Volumetric Studio (EVS) software package, which integrates the functionalities and tools of EVS Pro and MVS and adds some new features. The Earth Volumetric Studio offers more advanced volumetric gridding, geostatistical analysis, 4D visualization tools, and better performance and flexibility. Its graphical user interface (GUI) and modular analysis and graphics routines and tools can be customized and combined to suit the special analysis and visualization needs of any application. The commonly used features of the Earth Volumetric Studio are listed below.

- Borehole and sample posting
- Parameter estimation, using expert system driven 2D and 3D Kriging algorithms with best fit variograms
- Exploding geologic layers
- Finite difference and finite element modeling grid generation
- Advanced gridding
- High-level animation support
- Interactive 3D fence diagrams
- Multiple analyte data analysis and integrated volumetrics and mass calculation for soil and groundwater contamination and ore bodies
- 3D fault block generation
- Tunnel cutting
- Advanced texture mapping
- Mine pit modeling
- Visualizing and modeling of ore body overburden
- DrillGuide technology for quantitative appraisal of the quality of site assessments and identification of optimal new sample locations
- Comprehensive Python scripting of virtually all functions.

Ctech's EnterVol for ArcGIS is a suite of four extensions developed for ESRI's ArcGIS Desktop applications of ArcMap and ArcScene. Following is a list of these four EnterVol for ArcGIS extensions, which bring 3D volumetric modeling functionalities into the ArcGIS Desktop environment:

- EnterVol Volume Analyst: This extension is the analysis and visualization component of the EnterVol product suite. EnterVol Volume Analyst brings 3D modeling (such as 3D stratigraphic models, contaminated subsurface soil masses, groundwater plumes), subsetting (such as cuts, slices, cross sections), and volumetric analysis capabilities inside the ArcMap environment. Three-dimensional visualization of these models is performed inside ArcScene environment.

- EnterVol GeoStats: This extension brings the capability of building volumetric grids and performing parameter estimation into volume analysis in ArcGIS, with the assistance of the expert systems in performing 2D and 3D volumetric estimation using Kriging and inverse distance weighted methods.

- EnterVol Geology: This extension brings into ArcGIS the capability of constructing complex 3D volumetric geologic models, integrating lithology data with surfaces and computing volumes of each geologic layer or structure.

- EnterVol Tools: This extension brings several unique capabilities to ArcMap and ArcScene, such as building closed multipatches (pseudo volumes) from two sets of raster with any resolution/extent and adding view controls, labeled axes, data labeling, and direction indicators.

RockWorks

RockWorks is RockWare's flagship software product designed for subsurface data modeling and visualization, with tools such as maps, logs, cross sections, fence diagrams, solid models, and volumetrics. It is widely used in the petroleum, environmental, geotechnical, and mining industries. RockWorks supports a variety of data types, such as stratigraphy, lithology, quantitative data, color intervals, fracture data, hydrology, and aquifer data.

To better visualize the geology conditions of a project site, its lithology data can be interpreted and displayed with RockWorks in a variety of ways, including 2D geology maps created from intersection of subsurface lithology solid model and a horizontal plane (e.g., either the ground surface or a specified elevation surface), 3D lithology/boring logs, 3D solid stratigraphic models, geology cross sections, or fence diagrams. During a geotechnical investigation in an environmentally impacted site, geophysical, geochemical, and geotechnical data collected from the downholes can be analyzed, modeled, and visualized onsite with RockWorks to assist with making real time decisions. The data can be presented as downhole data logs, data cross sections, fence diagrams, continuous model isosurfaces, or 3D solid models that can then be cut into plan-view shaded contour maps at any specified elevation level. Aquifer and hydrology data of a project site can be analyzed and displayed as 2D contour maps of groundwater tables, time-based groundwater levels in boring logs/wells, 3D aquifer models, vertical ground-water profiles, or multipanel cross sections of groundwater and stratigraphy. RockWorks can also be used to analyze and visualize the fracture patterns of a project site. Fracture data can be displayed as oriented discs of varying colors and sizes on 3D logs, as fracture profiles/cross sections, fence diagrams, fracture isosurfaces, 3D voxel solid models, or as plan-view fracture data maps sliced from the 3D solid models at any given elevation levels.

With EarthApps tools, data can be extracted from the RockWorks data-sheet to generate various maps and flyovers for display in Google Earth. RockWorks supports SQL Server database, and a wide range of data formats through its import and export tools. Some of the data formats supported by RockWorks are listed below.

- Microsoft Excel
- LogPlot DAT
- LAS
- ASCII
- DBF
- DXF contour line
- DXF XYZ points
- Garmin TXT
- Geonics EM38
- GPL (Garmin GPS)
- Laser Atlanta survey files (ASCII and Trimble Pro XL)
- RockBase fixed-field
- SEG-P1 shotpoint data
- Spectrum Penetrometer
- Delorme GPL
- GSM-19
- MODPATH particle flowpaths
- Tobin WCS
- NEIC earthquakes
- Colog
- GDS and GDSII
- IHS PI/Dwight (well location, stratigraphy tops and perforation intervals)
- gINT (with gINT being installed, and a gINT correspondence file being created)

- Kansas Geological Survey data
- Newmont Assay MDB
- LidarXYZ
- Surfer Color Fill Table (*.lvl).

RockWare GIS Link is a software extension designed for ESRI's ArcGIS 10.x. It integrates some capabilities of RockWorks into ArcMap. With the RockWare GIS Link extension, GIS users can generate RockWorks style cross sections, profiles, fence diagrams, striplogs, and isopach maps within the ArcMap environment. It is worth noting that RockWare GIS Link is not a stand-alone software product because it needs the support of the RockWorks engine. Therefore, RockWorks has to be installed too. When both RockWorks and RockWare GIS Link are installed, RockWorks boreholes and/or wells data can be imported into an ArcGIS geodatabase and then into ArcMap as a point data layer. Cross sections, profiles or fence diagrams can be created by selecting the dedicated tool and then drawing line(s) to connect the selected boreholes and/or wells in ArcMap. Striplogs can be generated from the boreholes or wells by clicking on their locations in ArcMap with the striplog tool. Geology and other isopach or contour maps can be created in the DXF (AutoCAD's Drawing eXchange Format) format, with the isopath or contour tools.

Case Study 5.1: Modeling and Visualizing a Contaminated Groundwater Plume in NASB

As discussed earlier, due to the past ordnance and fuel-related activities in the formerly used Naval Air Station Brunswick (NASB) in the state of Maine between 1943 and 2011, especially during the World War II and the Cold War eras, soil and groundwater were impacted in many areas. A long-term groundwater monitoring and sampling program was conducted on the contaminated groundwater body, named "Eastern Plume," from 1995 to the 2000s, with roughly two to four sampling events per year. Figure 5.1 is a 3D topographic map showing the layout of the formerly used naval air station and the location of the Eastern Plume, which is represented by the pink polygon.

A large volume of data related to site geology and chemical conditions was collected. The Eastern Plume area at the formerly used NASB was classified as a Superfund site by the U.S. Environmental Protection Agency (EPA). Over 40 monitoring wells were constructed to form the monitoring and sampling network to monitor the changing conditions within the sand groundwater aquifer impacted by dissolved-phase volatile organic compounds (VOCs),

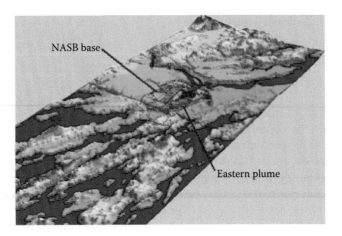

FIGURE 5.1
Three-dimensional topographic map of the formerly used NASB and its adjacent areas, created from the Digital Elevation Model data of the United States Geological Survey with GIS 3D modeling and analysis techniques. North is toward the upper right. Since the contaminated groundwater plume (the pink area) is located east of the NASB, it was named the Eastern Plume. There is a river north of the NASB and the ocean is south of it.

including TCE, PCE, 1,1,1-TCA and DCE. A pumping and treatment system was installed to remove groundwater contaminants and maintain hydraulic control of the VOC-contaminated plume.

In order to design effective remedial approaches, decision-makers required better tools to understand the geologic and hydrologic conditions of the impacted Eastern Plume area. Major concerns include the overburden geologic units, contaminant distribution within the geologic units, surface and groundwater flow patterns, and the interaction between groundwater and surface water. More than 200 borings were drilled to investigate the geologic and hydrologic conditions. GIS was integrated with other software applications, such as EquIS, GMS, and RockWorks, to model and visualize the geology, hydrology, and contaminated groundwater plume datasets. Based on the geologic data from the boring logs, a variety of geologic cross sections, fence diagrams, and 3D stratigraphy were generated. Figure 5.2 is a north-to-south geologic cross-section of the Eastern Plume area constructed from the geologic data interpreted from the soil boring logs and monitoring wells, with GIS and RockWorks techniques.

Figure 5.3 is a 3D model of the geologic strata (also known as stratigraphy) of the Eastern Plume area, generated from the geologic data interpreted from soil boring logs and monitoring wells, with the 3D Grid Module and other subsurface modeling tools of the Groundwater Modeling System (GMS). The 3D geologic data stored in a GIS database were geo-statistically analyzed and modeled using Kriging, inverse distance weighted (IDW), and Natural Neighbor interpolation methods within the GMS platform.

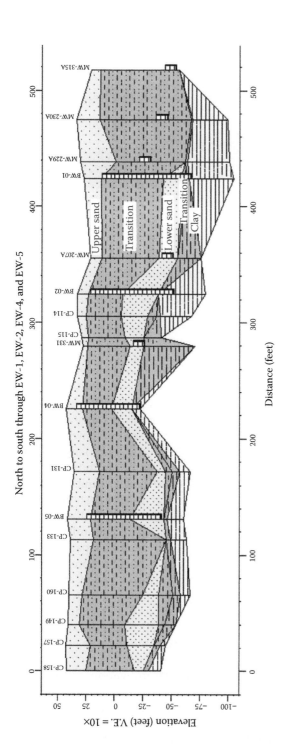

North to south through EW-1, EW-2, EW-4, and EW-5

Distance (feet)

Elevation (feet) V.E. = 10×

FIGURE 5.2

A north to south geologic cross-section of the Eastern Plume area at NASB. This cross-section also illustrates monitoring well screen intervals and their relationship to site geologic units. It shows four major geologic units: Upper Sand, Transition, Lower Sand, and Clay. The Lower Sand is the major layer containing and transporting the VOC-contaminated groundwater of the Eastern Plume. It thins out toward the northern and southern sides of the site and is confined between two thick lower-permeable units, i.e., the Transition unit above it and the Clay unit under it. The Clay unit contains the boring logs confirmed (gray) and inferred (white) layers. There are two depressions on the top surface of the Clay unit, one in the north side and one in south side of the site. Therefore, the migration of the VOC-groundwater plume laterally outside the site or downward toward the bedrocks was limited. With the VOC plume trapped mainly inside these two depressions, it made the remediation of the plume a little easier.

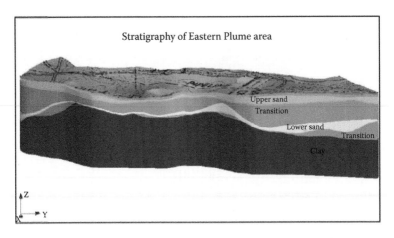

FIGURE 5.3
Three-dimensional stratigraphy of the Eastern Plume area at NASB. An USGS 7.5 minute topographic quadrangle map was draped on the top of the terrain to enhance visualization. With references to the features on the ground, it is much easier for users to study the underground geologic condition. North is on the right side and south is on the left side. This 3D model can be rotated and tilted to view the geologic units from any angles and perspective.

Both the geologic cross-section and this 3D stratigraphy illustrate that there are four major geologic units (also called layers): the Upper Sand, Transition, Lower Sand, and Clay. The Clay unit is divided into two layers, that is, the top Clay layer (shaded with gray in the geologic cross-section shown in Figure 5.2), which was confirmed by the soil borings and monitoring wells, and the bottom Clay layer (shaded with white in the geologic cross-section shown in Figure 5.2), which was interpreted based on the general geologic conditions of the project site area and statistical analysis of the geologic data from the soil boring logs and wells. The thickness and geometry of these geologic units, especially the Lower Sand and the Clay, change dramatically across the Eastern Plume area.

This 3D stratigraphy is very useful for geologists to study the geologic conditions of the project site, such as the dominant geologic units/layers, their relationships to each other, their thickness and geometry patterns, groundwater aquifer(s) and other related information. Furthermore, from this 3D stratigraphy, any individual geologic unit can be extracted and studied in more detail, without other units blocking the views of its top and bottom surfaces. For some special geologic units, such as high permeable sand and low permeable clay layers, the variations of their thickness and top surface geometry are crucial in understanding how contaminants are transported underground both laterally and vertically, and therefore they are important in designing the most accurate and effective mitigation and/or remedial strategies and approaches.

For example, the Lower Sand unit was extracted from the 3D stratigraphy to be examined in more detail, as shown in Figure 5.4. Since the Lower Sand unit is the major groundwater aquifer of the Eastern Plume area and

3D Lower Sand layer
(viewpoint from southeast)

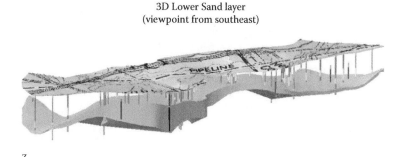

FIGURE 5.4

A 3D solid model of the Lower Sand geologic layer of the Eastern Plume area at NASB. The color bars represent the boring logs and wells used to generate this 3D geologic model. An USGS 7.5 minute topographic quadrangle map was draped on the top of the terrain to help users better visualize this 3D model. North is on the right side and south is on the left side. This 3D geologic model can be rotated and tilted to view this Lower Sand unit from any angles and perspective.

contains the VOCs-contaminated plume, it is a very important geologic unit that should be studied thoroughly in order to better understand the plume. With other geologic units not showing, it is now much easier to visualize the thickness and geometry of this Lower Sand unit in 3D space by rotating, tilting, and zooming in on this individual 3D geologic unit model. With a USGS 7.5 minute topographic quadrangle map draped on top of the terrain above this 3D Lower Sand unit model, it is much easier to relate underground geologic conditions to the features on the surface, such as the environmentally impacted area, possible sources of contaminants, surface water, properties, roads, and so on. This 3D model indicates that the thickness of the Lower Sand unit changes significantly throughout the site.

Another very important geologic unit is the Clay layer underneath the Lower Sand. Figure 5.5 is a 3D model of the top surface of the Clay unit. The deepest geologic layer the soil borings and wells penetrated into (approximately 80 feet below the surface) is the low permeable Clay unit, which serves as a natural barrier to prevent the vertical migration of the contaminated plume down into the bedrocks. In addition, the changes in clay thickness and the geometry of its top surface affect lateral groundwater flow and the transporting of the VOC-contaminated groundwater plume. Therefore, the 3D topography of the clay's top surface is very important to effectively understand the dynamics of the contaminated groundwater plume. Figure 5.5 shows a 3D view of the top surface of the underlying Clay unit and the outlines (in yellow) of the highest VOC concentrations inside the two depressions on the clay's top surface. This 3D surface model illustrates very

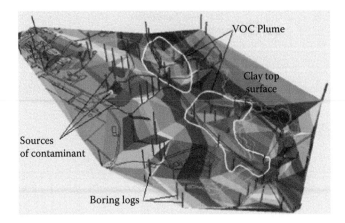

FIGURE 5.5

A 3D GIS model of the top surface of the underlying Clay unit, containing two naturally formed depressions (highlighted by yellow polygons), one in the north (top of the map) and one in the south of the site (bottom of the map), in the Eastern Plume area, NASB. The green colors represent lower elevation (forming the two depressions), while white or gray colors represent higher elevation and brown and dark yellow colors represent medium elevations. The highest concentrations of the total VOC-contaminated groundwater plume were confined inside the two depressions of the underlying Clay unit, which prevent the contaminants in the groundwater from migrating either outside the site laterally or downward into the underlying bedrocks vertically due to the low permeability property of clay.

well the spatial relationship between the contaminated groundwater plume and the topography of the top surface of the Clay unit.

These 3D graphics significantly helped geologists and environmental scientists to understand the conditions of the four major geologic units in the Eastern Plume area: the Upper Sand, Transition, Lower Sand, and Clay. The Lower Sand aquifer contained the majority of VOC-impacted ground water at this site. Total VOC concentrations up to 15,000 parts per billion were detected in it. The Lower Sand unit was the focus of remedial efforts at the site. Therefore, the thickness and geometric variations in three-dimensional space of this unit (as shown in Figure 5.4) are of particular importance.

The geologic cross-section (as shown in Figure 5.2) and the 3D stratigraphy models (as shown in Figure 5.3) demonstrate that the Lower Sand layer thins toward the northern and southern sides of the Eastern Plume area, and it is confined between two lower permeable units, that is, the overlying Transition unit and the underlying Clay unit. This relationship of geologic units indicates that the contaminated groundwater plume, which is contained mainly inside the Lower Sand layer, was confined inside the Eastern Plume area by the overlying Transition unit and the underlying Clay unit, with little potential of migrating outside of the project site. The 3D geometry of the Lower Sand layer was also useful in understanding the spatial distribution of this groundwater VOC plume. It appears that the highest concentrations of total VOCs were inside the north and south areas of the site where the underlying

Clay unit forms two depressions, as shown clearly on the 3D model of the top surface of the Clay unit in Figure 5.5 and the north-to-south geologic cross-section in Figure 5.2.

Groundwater elevation measurement data of the 40 water wells collected during the multiple sampling and monitoring events were statistically analyzed to generate groundwater elevation contours and interpret ground-water flow patterns and trends, as discussed in more detail earlier in the data analysis chapter. The groundwater elevation data were also used to con-struct 3D models of the groundwater table and flow patterns of the Eastern Plume and adjacent areas, as shown in Figure 5.6. The 3D groundwater table surface corresponds well with the topography of the top surface of the Clay unit, as shown in Figure 5.5, showing the two depressions, one in the north side and another one in the south side of the Eastern Plume area. The ground-water table and flow pattern changed during the sampling and monitoring events, therefore influencing the changes of the groundwater VOC plume too. Compared with the 2D groundwater elevation contour maps discussed earlier, 3D groundwater table surface models make it much easier to visual-ize the changes of groundwater conditions and their relationships with the changing patterns and trends of the groundwater VOC plume.

As Figure 5.1 shows, this formerly used naval air station is bordered in the north by a river and in the south by the ocean. The concern was that surface water might also be contaminated by the contaminants in the soil and groundwater. Therefore, surface water conditions and the interactions between groundwater and surface water were also investigated. Surface water flow patterns and drainage are mainly determined by the topography

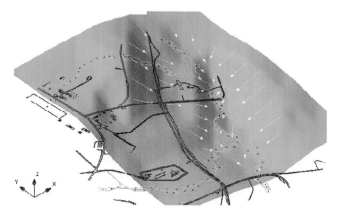

FIGURE 5.6

A 3D model of the groundwater table surface with two depressions, one in the north (top of the map) and another one in the south of the site (bottom of the map), in the Eastern Plume area of the NASB. The yellow arrows represent interpreted groundwater flow directions. The highest concentrations of the VOC-contaminated groundwater plume were mainly inside the two depressions. The light blue lines represent the surface water stream and its tributaries in the Eastern Plume area.

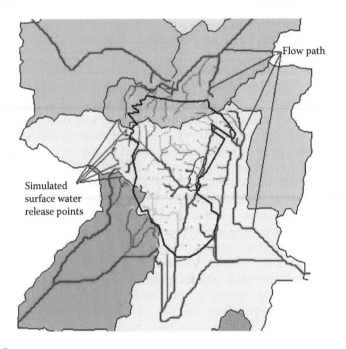

FIGURE 5.7
A GIS map showing the simulated surface water flow network and drainage basins of the formerly used Naval Air Station Brunswick (NASB), generated from the Digital Elevation Model (DEM) datasets of the United States Geological Survey (USGS). The surface water flow network inside the site was simulated using a particle tracking technique, with the green dots representing the imaginary particles released on the site randomly and the pink lines representing surface water flow paths from the locations of the released particles. The thick blue line is the NASB boundary and the yellow area with a black outline is the Eastern Plume. The other color polygons are drainage basins calculated from the USGS DEM data.

of an area. Using the Digital Elevation Model (DEM) data of the United States Geological Survey (USGS), a surface water flow network within the project site was simulated using the GIS particle tracking technique. Imaginary particles were randomly released from within the Eastern Plume area to simulate and visualize the flow paths both inside and outside the site, shown in Figure 5.7 as green dots and pink lines. Drainage basins were also delineated for the site and the surrounding areas from the USGS DEM data, also shown in Figure 5.7 as colored polygons. This simulated surface water flow network map indicates that the surface water from the NASB site could potentially flow toward the north and south of the site.

From 1995 to 2000, approximately 17 sampling events were conducted in the Eastern Plume area. A huge amount of chemical analytical data (with over 250,000 data records) were accumulated during this long-term monitoring and sampling program. Correctly understanding and interpreting these datasets was crucial for scientists, engineer and decision-makers to precisely understand the changes of the groundwater plume, to assess the pumping and treatment system and to determine future sampling and remedial strategies. These strategies included increasing or decreasing sampling frequencies, constructing new wells in places where the plume migrated to, decommissioning the wells where the contaminants had been removed and confirmed by multiple sampling events, and adjusting the pumping and treatment system.

Three-dimensional modeling and visualization techniques were utilized extensively to analyze the large and complex geologic and chemical analytical datasets to assist in understanding the changing patterns of the groundwater VOC-contaminated plume and make educated decisions. At least one 3D model was generated to represent the conditions of the VOC-contaminated groundwater plume at each sampling event. A few example 3D VOC plume models are shown in Figures 5.8 through 5.11. A 3D plume model could be rotated, tilted, and zoomed in to so that it could be viewed from all angles and perspectives to give a better understanding of the flume's shape and size at the specific sampling event. It could show different VOC concentration levels as well as calculate the plume volume at that

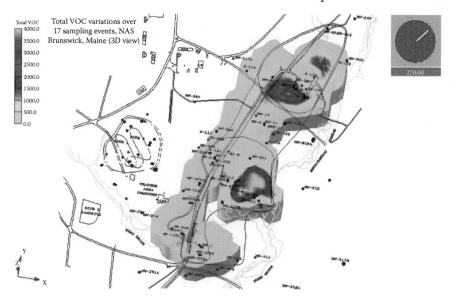

FIGURE 5.8
A snapshot of the animation showing the VOC concentration (in ppb) changes over the 17 sampling events from 1995 to 2000. The black points are water wells. The light blue lines represent streams. The purple, red, and yellow colors of the 3D plume model represent the highest total VOC concentration areas, one in the north side and another in the south side of the site. The blue and green colors indicate lower VOC concentrations. The clock with a number under it represents the time of the image frame (i.e., the 3D VOC plume model) in the 17 sampling events. The number represents days from the start of the long-term monitoring and sampling program. Since the sampling frequency was around 3–4 times annually, the interval of the samplings events was approximately 90–120 days. So, this image frame represents the 3D shape of the Eastern Plume in sampling event No. 3.

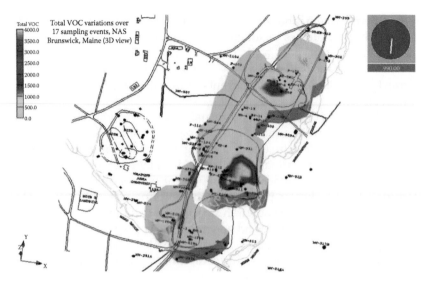

FIGURE 5.9
A snapshot of the animation showing the VOC concentration (in ppb) changes over the 17 sampling events from 1995 to 2000. This animation image frame represents the 3D shape of the Eastern Plume in sampling event No. 9. Compared with the 3D model of sampling event No. 3, this VOC plume became smaller overall, and the highest total VOC concentration areas contracted too, especially the north side one due to the pumping and treatment system.

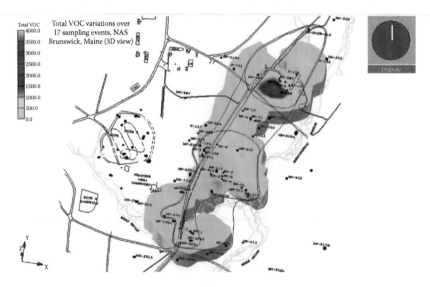

FIGURE 5.10
A snapshot of the animation showing the VOC concentration (in ppb) changes over the 17 sampling events from 1995 to 2000. This animation image frame represents the 3D shape of the Eastern Plume in sampling event No. 17. Compared with the 3D plume models of sampling events No. 3 and No. 9, the highest total VOC concentration areas contracted further, especially the south side one.

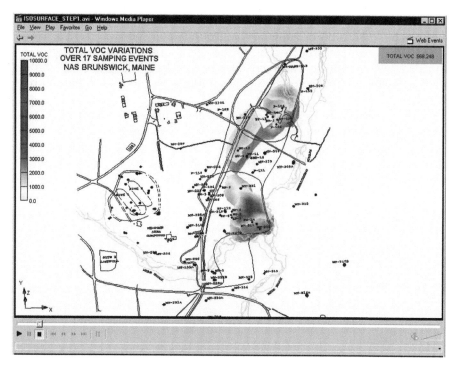

FIGURE 5.11

A 3D VOC plume model of sampling event No. 8, at the total VOC concentration level of 568 ppb or above. The red and purple colors of the 3D plume model represent the highest total VOC concentration areas, one in the north side and another in the south side of the site. This 3D model can show the plume at any concentration level and calculate its volume. The black points are water wells. The light blue lines represent streams.

concentration level, too. An animation was also created from the 3D VOC plume models of all 17 sampling events, which could be played at any frame rate and zoomed at any level. The 3D models and the animation of the VOC-contaminated groundwater plume also show clearly that the highest total VOCs concentrations (represented by purple, red, and yellow) were inside the north side and south side of the Eastern Plume area, in consistence with the geologic and groundwater analysis results, as discussed earlier.

After examining the individual 3D VOC plume models of the 17 sampling events and the animation film, scientist, engineers, and decision-makers had a much better understanding of the VOC concentration changing patterns and trends, which indicated that the pumping and treatment system worked well and the highest total VOC concentration areas (i.e., the purple, red, and yellow areas) were consistently reduced. At the end of the sampling event 17, the VOC concentration in the south of the Eastern Plume was significantly reduced. Therefore, future sampling and remedial actions should be focused more on the north side of the Eastern Plume area. Some wells in the south could be decommissioned or sampled less frequently, while a few more wells should be constructed in the north side of the site to catch and remove the VOC contaminants in the groundwater more efficiently. These sound, educated decisions increased the speed of the remediation process and also saved money by adding needed new wells while decommissioning wells no longer required and reducing sampling frequencies in the south of the site.

For this complex environmental site, 3D modeling and visualization techniques enhanced understanding and interpretation of geological, hydrological, and analytical data. Cost savings were achieved by using existing data to visualize site conditions, thereby maximizing the use of existing data while minimizing the need to collect additional data. The 3D models and visualizations of site geologic conditions and temporal and spatial changes in plume behavior were very useful in discovering relationships between geologic and chemical datasets and optimizing sampling and remedial decisions.

Case Study 5.2: Modeling and Visualizing a Contaminated Groundwater Plume at an Industrial Site in Bloomfield, Connecticut

Between 1999 and 2000, an environmental impact assessment project was conducted at an industrial site in Bloomfield, Connecticut. Its primary scope of work was to investigate the suspected dissolved-phase chlorinated volatile organic compounds (CHL-VOCs) in the groundwater and the geologic conditions of the site. Due to the sensitivity of the project site, its name, location, and other specific details cannot be discussed. Since this was a small project with a limited budget, tight schedule, restricted access and work space, and a preliminary study indicating possible VOC contamination in a shallow groundwater aquifer, direct push technology (DPT) was selected as the primary investigation and sampling method.

Direct push technology is a valuable alternative drilling, measuring, and sampling tool (in addition to heavy duty conventional drilling rigs) for environmental, geologic, and hydrological investigations. It has become popular since the 1990s in response to the growing need to assess sites more quickly and cheaply. Direct push allows rapid, cost-effective sampling and data collection from unconsolidated soils and sediments, typically down to a depth of 100 feet or more depending on the geologic condition of the site. A direct push device utilizes both hydraulic powered static force and percussion to drive or push small diameter steel boring rods (usually 4 in. or less in diameter) into the subsurface. The steel rods can also be equipped with various tools that provide continuous in situ measurements of subsurface properties such as geotechnical characteristics (e.g., penetrometer testing), geophysical properties (e.g., conductivity), and contaminant distribution (e.g., membrane interface probe). Soil, soil-gas, and groundwater samples can also be collected.

Compared with large and heavy conventional hollow stem auger drilling rigs, DPT has the following advantages:

- With smaller size and weight, and higher mobility, DPT allows drilling, sampling, and geophysical, geochemical, and geotechnical measuring in confined small areas that would be impossible for larger conventional hollow stem auger rigs.

- DPT's higher ground penetration rate makes data collection and site investigation more efficient and cost-effective than conventional drilling.

- With undisturbed soil cores, DPT allows for more accurate three-dimensional geologic modeling of the site.

- Direct push drilling is capable of collecting depth-discrete groundwater samples to locate contaminated layers more accurately.

- For VOC-contaminated soil, direct push drilling allows soil samples to be retrieved in plastic liners to prevent VOCs from being exposed to the atmosphere prior to being transported to the laboratory.

- Direct push drilling produces fewer cuttings. Also, its smaller drilling holes require fewer materials for well installation and sealing.

- With a dual-tube sampling method DPT reduces drill spoil, which requires safe disposal and adds to project costs.

- A smaller amount of waste material is produced because no water or drilling mud is introduced during direct push boring, unlike traditional drilling methods. Purge water disposal volumes are smaller too, because the volume of water extracted during well development and purging is much less due to the smaller radius of the disturbed aquifer around the well.

Due to the small project budget, tight schedule, and limited work space of this project, the direct push drilling, measuring, and sampling method was the best choice to quickly, cost-effectively, and accurately investigate both the geologic and hydrological conditions and the groundwater VOCs contamination of this industrial site. During the direct push drilling, underground soil electric conductivity data were collected automatically, onsite geologic logging was performed, and groundwater samples were retrieved. GIS technology was integrated with other data processing, analysis, and modeling software applications, such as EquIS, GMS, EVS, and RockWorks, to model and visualize the geology, hydrology, and the CHL-VOC-contaminated groundwater data.

Electrical conductivity is the ability of a material to conduct or transmit an electrical current. It is commonly measured in units of millisiemens per meter (mS/m), or decisiemens per meter (dS/m), which is 100 times greater than millisiemens per meter. Two different types of sensors are commonly used to measure electrical conductivity, namely, contact sensors and noncontact sensors. As its name indicates, contact sensors have to contact soil to measure its electrical conductivity. Based on the electromagnetic induction method, noncontact electrical conductivity sensors do not have to contact soil directly to measure its electrical conductivity.

A material's electrical conductivity is determined by its physical and chemical properties. For example, with smaller soil particles, clay conducts more current than silt and sand, which contains larger particles. Therefore, electric conductivity measurements are commonly used to classify geologic units, as shown in Figure 5.12. This is a quick and low-cost method of studying a project site's geology. It corresponds well with the more accurate and detailed cross sections generated from the geology logs, as shown in Figure 5.13. From the geologic logs, 3D stratigraphy models also were constructed for this site, as shown in Figure 5.14.

Since the shallow sand and silt geology unit is the major shallow groundwater aquifer—which was contaminated by dissolved-phase total chlorinated volatile organic compounds, as confirmed by the chemical analytical results of the groundwater samples collected during the direct push drilling—its thickness and geometry information were very important in delineating the VOC-contaminated groundwater plume and evaluating its potential movements. The silt geologic layer under the sand and silt is also very important because it is a natural barrier for groundwater flow due to its low permeability. In order to help scientists, regulators, the site owner, and other relevant stakeholders to better visualize and understand the geologic conditions of the project site, assess the VOC impact on the groundwater, and evaluate current and potential future risks, 3D models of the sand and silt layer (Figure 5.15) and its underneath silt layer (Figure 5.16) were constructed. The 3D models of these important geologic layers could be rotated and tilted so they could be viewed from any angle or perspective. A recent aerial photo was draped on top of the ground surface of the models to help viewers better visualize the relationship of the geology layers and the industrial site layout on the ground.

Figures 5.15 and 5.14 illustrate that the thickness of the sand and silt layer deceases toward the east side of the site, while it increases toward the west side of the site. Therefore, more direct push borings (represented by the color bars) were planned for the west side of the site. The colors of the boring logs of the 3D solid model of the sand and silt layer and the 3D model of the silt layer are the same as the 3D stratigraphy model, that is, yellow representing the top sand layer, green representing the sand and silt layer, purple representing the silt layer and brown representing the interbedded layer. The 3D models of the stratigraphy (Figure 5.14) and the silt layer (Figure 5.16) show that there is a small depression on the top surface of the silt layer in the northwest corner of the side. This depression and the increased thickness of

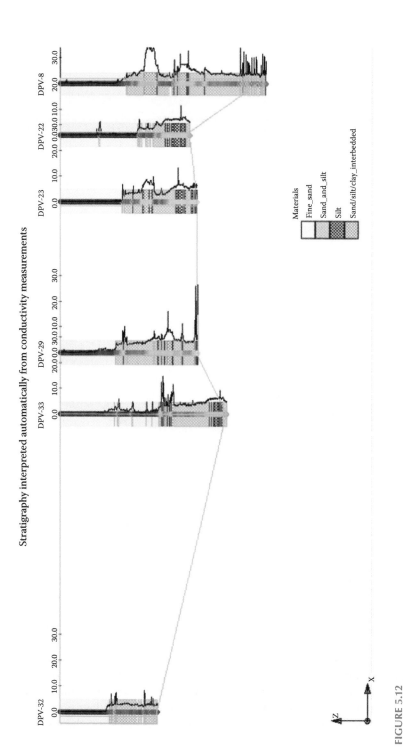

FIGURE 5.12

Geologic units interpreted from the electric conductivity measurements collected automatically during direct push drilling. Clay, with the finest particles, has the highest electric conductivity, while sand, with larger particles, has the lowest electric conductivity. Silt, with medium sized particles, has medium electric conductivity.

NW to SE (A–A') cross-section, Bloomfield

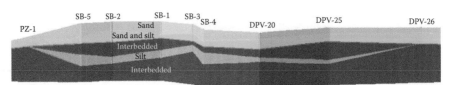

FIGURE 5.13
A geologic cross-section generated from the geologic logs compiled during direct push drilling. Chemical analytical results from groundwater samples indicated that a high concentration of total chlorinated volatile organic compounds (CHL-VOCs) existed in the shallow sand and silt layer. Since beneath it are the low permeable small-grained silt layer and the interbedded (consisting mainly of clay, silt, and sand) layer, the CHL-VOCs would not be likely to migrate downward into the bedrocks or deeper water aquifer(s). However, the CHL-VOCs contaminants could migrate laterally out of the site.

3D Stratigraphy of Bloomfield, viewpoint at southwest

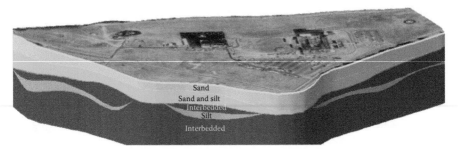

FIGURE 5.14
A 3D stratigraphy of the Bloomfield industrial site constructed from the geologic logs from the direct push drilling. A recent aerial photo was laid over the top of the terrain to enhance visualization so that underground geologic conditions could be related to the features on ground surface. This 3D model can be rotated, tilted, and zoomed to view the geologic units from any angle and perspective. North is toward the top side of this 3D model, east toward the right side, west toward the left side, and south toward the lower side.

the sand and silt layer in this sensitive area of the site elevated the possibility of a VOC-contaminated groundwater plume forming in the area, one that could spread laterally if not mitigated. Therefore, groundwater sampling should be focused more in the depression and its adjacent area.

However, the VOC plume should have very limited potential of moving downward into deeper aquifer(s) or bedrocks, due to the low permeability of the silt layer and the very thick interbedded layer, which consists mainly of clay, silt, and sand. Therefore, it was not necessary to perform deep drilling and sampling using conventional drilling methods at the preliminary environmental impact assessment investigation stage due to the low possibility of contamination in deep aquifer(s) and the bedrocks based on the geologic investigation.

Sand and silt layer, Bloomfield, viewpoint at south

FIGURE 5.15
A 3D model of the sand and silt geologic layer of the Bloomfield industrial site constructed from the geologic logs from the direct push drilling. The color bars are direct push boring logs, with yellow representing the top sand layer, green representing the sand and silt layer, purple representing the silt layer, and brown representing the interbedded layer. These are exactly the same as the colors on the 3D stratigraphy and boring logs in the cross-section. An aerial photo was draped on the top of the terrain to enhance visualization of the relationship of the geologic layer and the layout of the site on the ground surface. North is toward the top side of this 3D model, east toward the right side, west toward the left side, and south toward the lower side. This 3D model can be rotated, tilted, and zoomed to view the geologic layer from any angle and perspective. This model shows that the sand and silt is thicker in the northwestern corner of the site.

FIGURE 5.16
A 3D model of the silt geologic layer of the Bloomfield industrial site constructed from the geologic logs from the direct push drilling. An aerial photo was laid over the top of the terrain to enhance visualization of the relationship of the geologic layer and the layout of the site on the ground surface. North is toward the top side of this 3D model, east toward the right side, west toward the left side, and south toward the lower side. This 3D model can be rotated, tilted, and zoomed to view the geologic layer from any angles and perspective. This model shows that there is a small depression on the top surface of the silt layer in the northwest corner of the site.

Groundwater elevation measurements were statistically analyzed to generate groundwater elevation contours and interpret groundwater flow patterns, as shown in Figure 5.17. This GIS map shows that the groundwater of this industrial site flows mainly toward the east, northeast, and southeast. The VOC concentration data (or estimates) collected during direct push drilling using the membrane interface probe (MIP) tool were also statistically analyzed and displayed as color-shaded patterns on this groundwater contours map.

The MIP, also known as a semi permeable membrane sensor, is a quick and cheap screening tool for locating volatile organic compounds in the subsurface. During direct push drilling, a membrane interface probe tool is installed on the drilling rod to quickly measure VOC concentrations. While drilling,

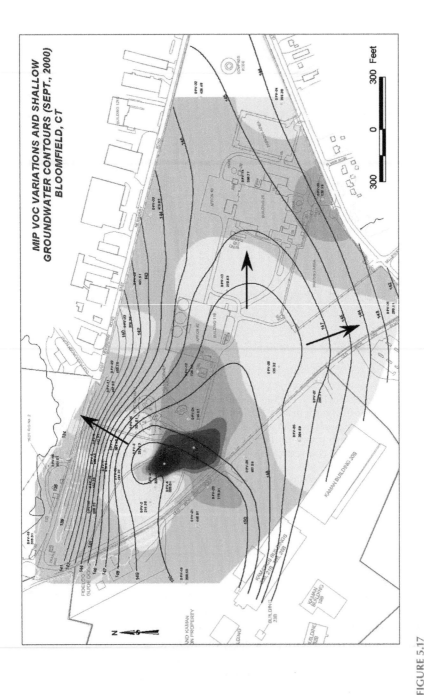

FIGURE 5.17

A GIS map showing shallow groundwater elevation contours and interpreted flow directions in September 2000. The blue lines are groundwater elevation contours. The black arrows are interpreted groundwater flow directions. The color-shaded patterns represent the estimated concentrations of the total chlorinated volatile organic compounds in shallow groundwater, analyzed from the direct push membrane interface probe (MIP) VOC measurements. The dark red and bright red colors represent the highest VOC concentration, while the light red color represents the lower VOC concentration.

the membrane interface probe heats the soil and groundwater adjacent to its tip, causing volatilization of any volatile organic compounds. The volatilized VOCs then diffuse across a thin permeable membrane on the probe's side. Once inside the probe, the VOCs are brought to the surface using a carrier gas through tubing which is connected to a laboratory-grade electron capture device (ECD), photo ionization detector (PID), flame ionization detector (FID), and other detectors for immediate analysis on site.

Figure 5.17 illustrates that the highest VOC concentration (represented in dark red) is in the west of the industrial site. Since the groundwater flows in general toward the east, the VOC-contaminated shallow groundwater plume could migrate toward east throughout the site and possibly out of the site as well.

The chemical analytical results from the groundwater samples indicate high concentrations of the dissolved-phase total CHL-VOCs in the groundwater, especially in the northwest corner of the site. Although it was clear that there was a VOC-contaminated groundwater plume in the site, by looking at the large quantity of the chemical analytical data, it was very hard for the environmental scientists, managers, regulators, site management, and other related stakeholders to locate the VOC plume and understand its extents in a 3D space under the ground. GIS and other 3D-modeling and visualization techniques were utilized to process the large volumes of the chemical analytical data, which were combined with the site geologic and hydrologic data, and to generate a 3D model of the VOC-contaminated groundwater plume. Figure 5.18 is a planar view of the 3D model of the VOC-contaminated

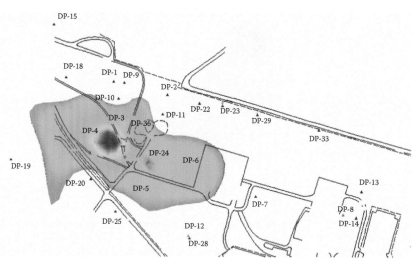

FIGURE 5.18

Planar view of the 3D model of the total chlorinated volatile organic compounds contaminated groundwater plume in the northwest corner of the site, near DP (direct push)-16, DP-24, DP-3, and DP-4 area. Red represents the highest VOC concentration, followed by yellow and green. Blue represents the lowest VOC concentration. North is toward the top side of this 3D model, east toward the right side, west toward the left side, and south toward the lower side.

groundwater plume. It demonstrates that the highest VOC concentration (the red area) was located in the northwest corner of the side near DP-4, DP-3 and DP-16, as was predicted based on the analysis of the geologic conditions. Red represents the highest VOC concentration, which is followed by yellow and green. The blue area represents the lowest VOC concentration. It also indicates that the VOC-contaminated groundwater plume could migrate laterally throughout and out of the site, corresponding well with the predictions from the geologic studies, too.

Figure 5.19 is a 3D view of the VOC-contaminated groundwater plume at approximately 2000 parts per billion (ppb) level or above. Similar to the planar view, this 3D model also indicates the VOC plume could spread laterally, especially toward the east of the site. The 3D VOC plume model could be cut into cross sections in any direction to visualize the VOC concentration changes inside it, as shown in Figures 5.20 and 5.21.

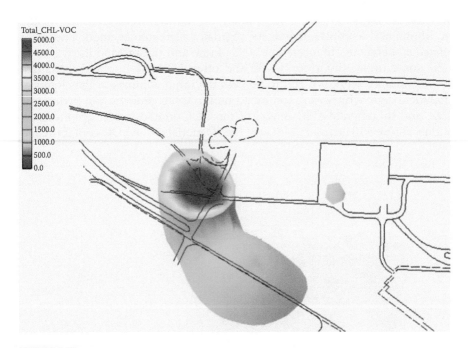

FIGURE 5.19

A 3D model of the total chlorinated volatile organic compounds contaminated groundwater plume in the northwest corner of the site. North is toward the top side of this 3D model, east toward the right side, west toward the left side, and south toward the lower side. The red represents the highest VOC concentration (in ppb), followed by yellow, green, and blue. Since this 3D model represents VOC concentration at around 2000 ppb level or above, blue is not shown in the model. This 3D plume model can show VOC concentration at any level and calculate its volume at that level. This 3D model can be rotated, tilted, and zoomed to view the VOC-contaminated groundwater plume from any angle and perspective. It can also be cut into cross sections to visualize VOC concentration changes inside this 3D model as shown on next three figures.

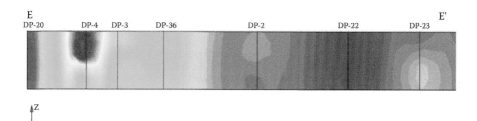

FIGURE 5.20

A VOC concentration cross-section sliced from the 3D VOC solid model along the SW-NE (south west–north east) line. It shows VOC concentration changes inside the 3D model along this direction. This VOC cross-section indicates that the highest VOC concentration was in the shallow groundwater aquifer in the northwest corner of the site, near DP-4, not migrating downward much. However, it had the potential of spreading laterally. It can show VOC concentration at any level. The purple and red represent the highest VOC concentration, followed by yellow and green. Blue means low VOC concentration.

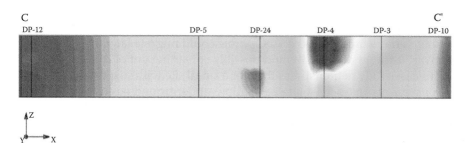

FIGURE 5.21

A VOC concentration cross-section sliced from the 3D VOC solid model along the NW-SE line. It shows VOC concentration changes inside the 3D model along this direction. This VOC cross-section also indicates that the highest VOC concentration was in the shallow ground-water aquifer in the northwest corner of the site, near DP-4, not migrating downward much. However, it had potential of spreading laterally. It can show VOC concentration at any level. The red and purple represent the highest VOC concentration, followed by yellow and green. Blue means low VOC concentration.

The southwest to northeast (SW–NE) cross-section of the VOC plume shown in Figure 5.20 indicates that the highest VOC concentration, represented by the purple and red areas was in the shallow aquifer and it had the potential of moving to the east side of the site if no mitigation or remediation measures were taken. The northwest to southeast (NW–SE) cross-section of the VOC plume shown in Figure 5.21 demonstrates similar patterns and treads of the VOC concentration changes in the site. VOC concentration values could also be contoured vertically, as shown in Figure 5.22, to further enhance the visualization and understanding of the VOC concentration patterns. The thick red contour lines represent higher VOC concentrations. It

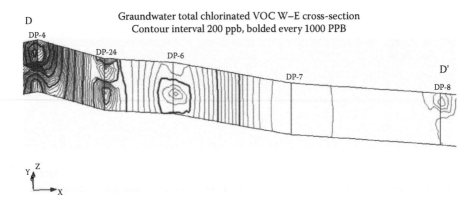

FIGURE 5.22

A VOC concentration contour cross-section sliced from the 3D VOC solid model along the west–east line. It shows VOC concentration changes inside the 3D model along this direction. This VOC cross-section also demonstrates that the VOC-contaminated groundwater plume had potential of spreading laterally.

further demonstrates that the VOC plume could migrate laterally, especially eastward throughout the site, and possibly spreading out of the site.

This case study demonstrates that, for a small site with suspected contaminations mainly in unconsolidated shallow underground (around 100 feet or so, or above), direct push technology can be used to quickly and cost-effectively collect geophysical, geochemical, geological, and hydrological data as well as to retrieve soil, soil-gas, and groundwater samples to characterize the site. GIS technology can be integrated with other software applications to quickly process, analyze, model, and visualize large quantities of datasets to help scientists, decision-makers and other relevant stakeholders to better visualize and understand the complex datasets; find patterns and trends; and make sound, educated decisions to mitigate and/or remediate the environmentally impacted site conditions.

Case Study 5.3: Modeling and Visualizing the Contaminated Soil at Site 2A, Inside the Former Camp Sibert

As discussed earlier, in preparing for World War II, Camp Sibert, located in northeast Alabama, was selected in the spring of 1942 for use in the development of a Replacement Training Center (RTC) for the U.S. Army Chemical Warfare Service. The RTC was moved from Edgewood, Maryland, to Alabama in the summer of 1942. In the fall of 1942, the Unit

Training Center (UTC) was added as a second command. Units and individual replacements were trained in aspects of basic military training and in the use of chemical weapons, including decontamination procedures and smoke operations from late 1942 to early 1945. Mustard, phosgene, and possibly other agents were used in the training. The facility provided the opportunity for large-scale training with chemical agents that had previously been unavailable. In addition to chemical training, several types and calibers of conventional weapons were fired at the former Camp Sibert, with the 4.2 in. mortar being the heavy weapon used most in the training. Firing ranges at the camp included those for .30 caliber rifle, machine gun, .22 caliber rifle, submachine gun, .45 caliber pistol, grenade (for both rifle and hand grenades), 4.2 in. mortar (used for training in the use of white phosphorous and high explosive rounds), artillery, bazookas, and antiaircraft.

Site 2A is located on what was tract C230 in the central portion of the former Camp Sibert. It was formerly used for chemical agent decontamination training and is known to be a burial site for training materials. Instruction at this site included decontamination training on an airplane fuselage, walls, floors, different road types (gravel, concrete, sand, and macadam), shell holes, and trucks. The site also included a mustard soak pit, a chemical agent (mustard) storage area, a supply building, and a shower/dressing building. Reports from previous investigations concerning Site 2A indicate that the chemical agents used for training or buried at Site 2A include mustard (H, HD), nitrogen-mustards (HN1 and HN3), and lewisite (L). Industrial chemicals such as phosgene (CG), fuming sulfuric acid (FS), tearing agent (CNB), and adamsite (DM) may have also been used.

When World War II ended in 1945, the former Camp Sibert was no longer needed for training, and the whole installation, including Site 2A, was closed. Some chemical training materials and equipment were buried in the burial pits, and excess chemical agents and industrial chemicals were dumped into the mustard soakage pit inside Site 2A. Therefore, the soil of Site 2A was contaminated. Figure 5.23 is a GIS map showing the three burial pits, various chemical agent decontamination training areas, and the mustard soakage pit. The mustard soakage pit, located in the south central area of Site 2A, is the focus of this case study.

During the 2009 chemical agent contaminated soil removal action (RA) and the remedial investigation/feasibility study (RI/FS) investigations at Site 2A of the former Camp Sibert, the soil in the mustard soakage pit was sampled for the chemical agents and agent breakdown products. Figure 5.24 shows the arsenic concentration values from the soil samples taken during the excavation phase of the RA project. The colored squares are the 10 × 10 feet excavations grids and the numbers are arsenic concentration results and their sampling depths (in parentheses), ranging from 1–20 feet.

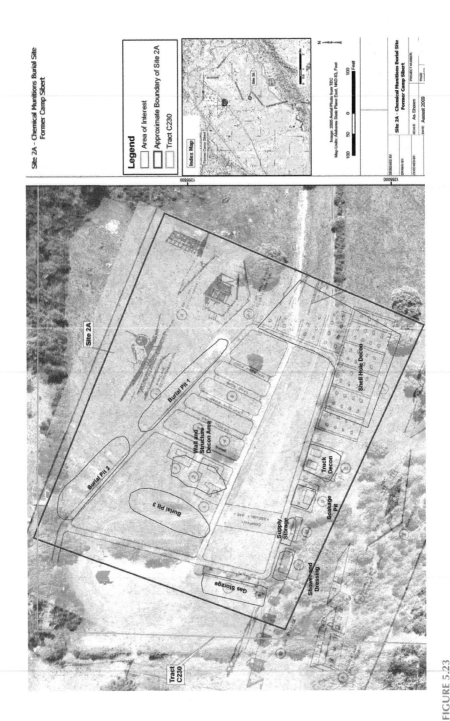

FIGURE 5.23

Areas (blue line polygons) impacted by former chemical agent decontamination training and burial activities from 1941 to 1945, at Site 2A (the red line polygon) of the former Camp Sibert. The mustard soakage pit is located in the south-central portion of Site 2A.

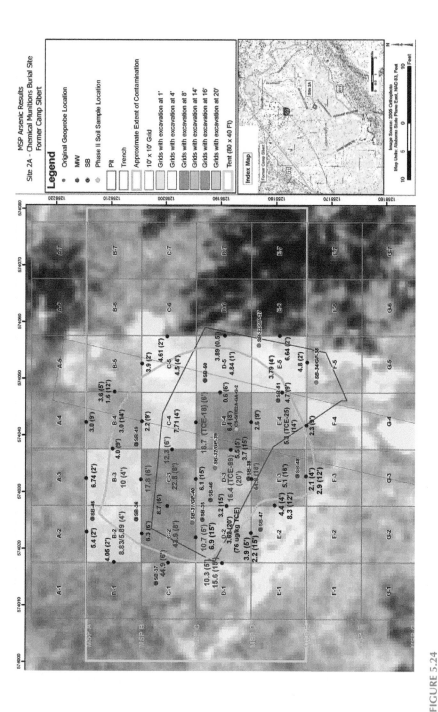

FIGURE 5.24

Arsenic concentration results in the soil of the mustard soakage pit at Site 2A in the former Camp Sibert. The red polygon is the mustard soakage pit. The color squares are 10 feet × 10 feet excavation grids, with the colors representing their excavation depth, ranging from 1 to 20 feet. The color points are samples. The numbers are arsenic concentrations and the sampling depths. The thick orange line is the estimated extent of the contamination on the ground surface, based on the sampling results.

The sampling results from the preliminary soil boring investigation were discussed in the data analysis chapter. Combining the soil sampling results from all phases of the investigations, a 3D model of the contaminated soil of the mustard soakage pit was developed, as shown in Figure 5.25. With this 3D model, it is much easier to visualize the contamination underground in a 3D space. It can be rotated, tilted, and zoomed in to view the contamination patterns from any angle and perspective. It can also show the contamination at any concentration level (Figures 5.26 and 5.27) and calculate the soil volume and the corresponding costs to excavate and transport it to a processing facility.

Figure 5.28 illustrates how data analysis, modeling, and visualization help in understanding complex data, identifying and delineating environmental contamination on the surface and underground, and in designing mitigation and/or remedial strategies.

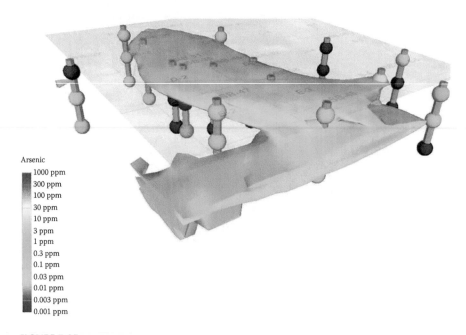

FIGURE 5.25

A 3D arsenic concentration (at 8 ppm or above level) model of the mustard soakage pit inside Site 2A of the former Camp Sibert. This 3D model can be rotated, tilted, and zoomed to view the contaminated soil body from any angle and perspective. The red represents the highest arsenic concentration, followed by orange, yellow, and green. The blue represents the lowest concentration. The bars with the color balls represent soil borings. The color balls represent concentrations at the generalized three depth intervals, i.e., shallow, intermediate, and deep. This 3D model indicates that the higher arsenic concentration was mainly in the area of SB (soil boring)-31, SB-35, and 3B-46. This model can display the contaminated soil body at any arsenic concentration level and calculate its corresponding volume/weight as well as estimate the cost of remediation.

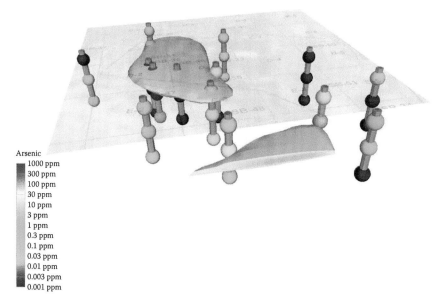

FIGURE 5.26

A 3D arsenic concentration (at 25 ppm or above level) model of the mustard soakage pit inside Site 2A of the former Camp Sibert. This 3D model can be rotated, tilted, and zoomed to view the contaminated soil body from any angle and perspective.

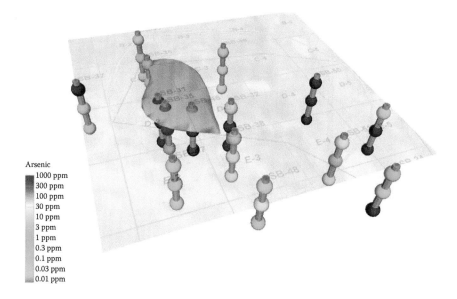

FIGURE 5.27

A 3D arsenic concentration (at 100 ppm or above level) model of the mustard soakage pit inside Site 2A of the former Camp Sibert. This 3D model can be rotated, tilted, and zoomed to view the contaminated soil body from any angle and perspective.

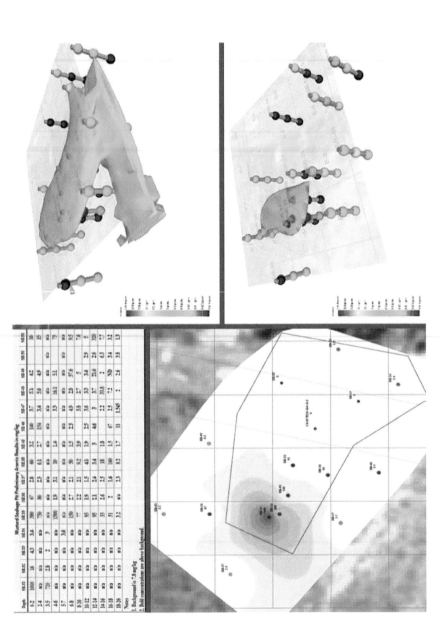

FIGURE 5.28
Arsenic concentration data table (top left), color-shaded arsenic concentration pattern of 0–2 feet depth interval (bottom left), 3D arsenic concentration model at 8 ppm level or above (top right), and 3D arsenic concentration model at 100 ppm level or above (bottom right).

The left side of Figure 5.28 contains a data table showing the results of the preliminary soil boring analysis and the statistically analyzed arsenic concentration patterns. As discussed in the data analysis chapter, a variety of color-shaded arsenic concentration maps were generated from the chemical analysis database, with one map for each sampling depth interval ranging from 0 to 20 feet, at 2-feet intervals. These two-dimensional maps were more helpful in understanding the site contamination conditions than the analytical numbers in the data table. But it was still hard for project scientists and decision-makers to visualize the relationships among the many 2D concentrations maps. It was also difficult to calculate the contaminated soil volumes at certain concentration levels. To overcome these challenges, GIS was integrated with other modeling software applications to generate 3D arsenic-contaminated soil models, as shown on the right side of Figures 5.28 and in more detail, in Figure 5.25 (at 8 ppm or above concentration level), Figure 5.26 (at 25 ppm or above concentration level), and Figure 6.27 (at 100 ppm or above concentration level). With these 3D arsenic concentration models, scientists and decision-makers could visualize the arsenic contamination situations more accurately and make better remedial decisions.

Case Study 5.4: Modeling and Visualizing the Contaminated Soil in Bronx Criminal Court Complex, New York City

The Bronx Criminal Court Complex is located in the northern part of New York City. An environmental investigation project was conducted between 2000 and 2002 to evaluate the soil and groundwater contaminated by volatile organic compounds (VOC), semi-volatile organic compounds (SVOC), pesticides, polychlorinated biphenyls (PCBs), and some metals as well as to investigate possible underground storage tanks (USTs) and recommend remedial options and calculate their costs. The project area was divided into three investigation sites, named Site A, Site B, and Site C, bounded by East 161st Street in the south, East 163rd Street in the north, Sherman Avenue in the west, and Morris Avenue in the east, as shown in Figure 5.29.

The project site geology was interpreted and modeled from soil boring and geotechnical data. GIS and other software applications, such as EquIS Geology, GMS, EVS, and RockWorks, were utilized to process the data and generate a variety of geologic cross sections, fence diagrams, and 3D stratigraphy models. Figure 5.30 is a 3D model of the strata (also known as stratigraphy) of Site A, Site B, and Site C. Figure 5.31 contains two geologic cross sections, one cutting through Site B and Site A in the south of the project area

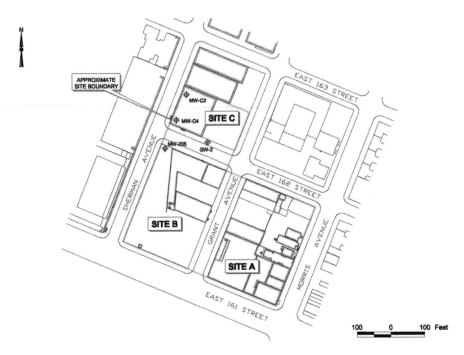

FIGURE 5.29

Layout of the Bronx Criminal Court Complex project site in northern New York City. The orange line is the boundary of the project area, which contains three investigation sites, Site A, Site B, and Site C. The red symbols represent the locations of the existing groundwater and monitoring wells.

and another one cutting through Site C in the north and Site A in the east. Figure 5.32 is a 3D surface model of the top of the bedrock.

As Figures 5.30 and 5.31 illustrate, there are four major geologic layers or units in the project sites. The top fill material layer is up to 25 feet thick, consisting mainly of sand, silt, brick, metal, and concrete. Beneath this top fill layer is a natural soil layer of up to 25 feet in thickness, lying on top of the weathered bedrock layer. The natural soil layer is comprised mainly of fine sand, clay, and silt. At Site C, a clay layer lies over the weathered bedrock. This is a very important geologic layer because petroleum release was confirmed during tank excavation sampling in Site C in June 2000. Due to its low permeability, this clay layer prevents the released petroleum from flowing downward into groundwater and bedrock. The weathered bedrock varies in thickness from 0 to –40 feet. The bedrock is mainly Inwood limestone and/or Fordham gneiss. As the 3D stratigraphy model and cross sections show, the thickness of these geologic layers varies dramatically across the project sites. The top surface of the bedrock has irregular topography too, with a valley in Site B and a ridge in Site A, as shown in Figure 5.32.

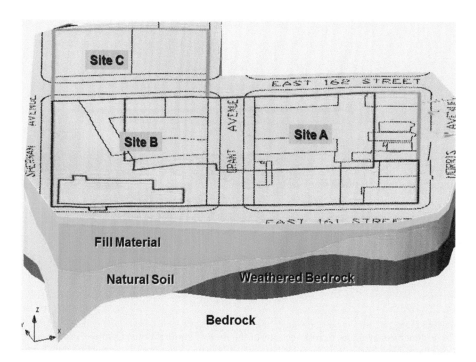

FIGURE 5.30

A 3D geology stratigraphy of the Bronx Criminal Court Complex project sites in northern New York City. It was put together based on the geology data from the soil borings and geotechnical studies. A site layout map was laid over the top of this 3D geology model to help readers better visualize the geologic conditions at the project site. The colors from top to bottom are green representing the top Fill Material layer, yellow representing the Natural Soil layer, and blue representing the Weathered Bedrock layer. Since the bedrock was not penetrated through, its thickness was unknown and not included in this model. This 3D geology stratigraphy model can also be rotated, tilted, and zoomed so that it can be studied from any angle, perspective, and distance.

More than 45 soil borings were drilled. From them, 43 environmental samples were collected and analyzed for VOC, SVOC, pesticides, PCBs, and metals. Two extraction wells, 2 observation wells and 17 monitoring wells (16 overburden wells and 1 bedrock well) were installed for pumping test and sampling. The water wells were constructed with 10–15′ screens. Groundwater level gauging was conducted on a weekly basis, and samples were collected on a quarterly basis for analysis of VOC, SVOC, pesticides, PCBs, and metals. A ground-penetrating radar (GPR) survey was conducted to investigate possible underground storage tanks (USTs). Due to the fill material, GPR penetration investigation was limited to a depth of 8–10 feet Figure 5.33 shows the locations of the soil boring, wells and underground storage tanks.

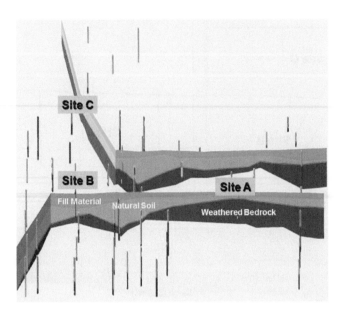

FIGURE 5.31
Three-dimensional geology cross sections of the Bronx Criminal Court Complex project sites in northern New York City. The color bars are soil borings. The front cross-section cuts from the southwest corner of Site B (lower left on this figure), going north inside Site B, and then cutting eastward through Site A (right side). The back cross-section cuts from the northwest corner of Site C (upper left), going south inside Site C, and then cutting eastward through Site A (right side). The color bars are soil boring logs. Their colors from top to bottom are green representing the top Fill Material layer, yellow representing the Natural Soil layer, blue representing the Weathered Bedrock layer, and red representing the Bedrock layer. With these two geologic cross sections, it is much easier to visualize the geologic conditions of the three project sites: Site A, Site B, and Site C.

Groundwater level gauging data were statistical analyzed using GIS geospatial analysis techniques, based on Kriging, inverse distance weighted (IDW), and other interpolation methods, to generate groundwater elevation contours and interpret flow directions, as shown in Figure 5.34. This map indicates that groundwater flows in general from north (Site C) to south (Site B), and east (Site A) to west (Site B). Groundwater sampling results were analyzed and displayed on maps, with one map generated for each sampling event.

Figure 5.35 is a data map of sampling event No. 3, showing the concentrations of VOC, SVOC, and metals in the groundwater samples. This map demonstrates that groundwater samples in Site B and the western portion of Site A contained higher concentrations of VOC, SVOC, and metals, while samples in the northern Site C (i.e., C2 and C4) and eastern Site A (i.e., MW-251) had less concentrations of VOC and SVOC. Also, no metals were detected in the samples taken from northern Site C and eastern Site A either. These

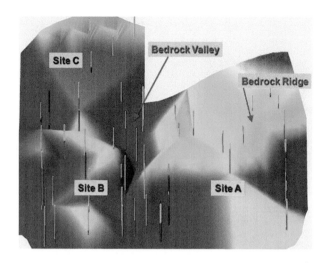

FIGURE 5.32

Three-dimensional top surface of the bedrock of the Bronx Criminal Court Complex project site in northern New York City. The color bars are soil boring logs. The colors of the surface represent elevations. Red indicates the highest elevation, followed by yellow, green, and cyan. Blue represents the lowest elevation.

groundwater sampling results correspond well with the groundwater flow patterns of the project sites.

Similar sampling results data maps were generated for soil samples. Figures 5.36 through 5.39 are four example soil sample analytical results maps. Figure 5.36 shows the analytical results of pesticides and PCBs in Site B soil samples collected in August 2000. Figure 5.37 displays the concentrations of volatile organic compounds (VOCs) in soil samples collected from Site A in August 2000. Figure 5.38 illustrates the concentrations of semi-volatile organic compounds (SVOCs) in Site B soil samples collected in August 2000. Figure 5.39 shows the concentrations of metals in Site B soil samples collected in August 2000. These maps indicate that soil in Site B and Site A was contaminated by VOCs, SVOCs, metals, and pesticides/PCBs.

A grid system of 25 feet x 25 feet horizontal spacing was designed to systematically sample, delineate, and characterize the contaminated soil in Site A and Site B, as shown in Figure 5.40. In each grid, soil samples were collected from three depth intervals, that is, 0–5 feet, 5–10 feet, and 10–15 feet, respectively, to delineate and characterize impacted soil bodies in a 3D space. GIS 3D modeling techniques were utilized to statistically analyze the sampling results and construct 3D impacted soil models. For this specific project, soil was broadly characterized into three categories: hazardous soil, petroleum-impacted soil, and nonhazardous soil. Their volumes were calculated from the 3D models. Four soil treatment or disposal options were proposed, namely, biological

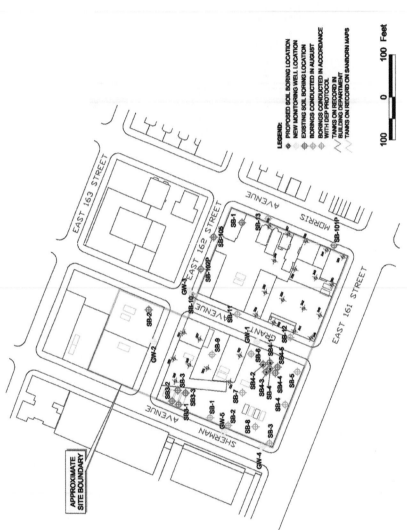

FIGURE 5.33

A GIS map showing locations of the soil borings, water wells, and underground storage tanks (USTs) inside the Bronx Criminal Court Complex project sites in northern New York City.

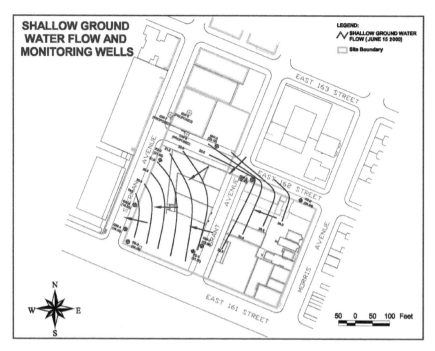

FIGURE 5.34
An example GIS map showing groundwater elevation contours and flow directions within the Bronx Criminal Court Complex project sites in northern New York City. In general, groundwater flows from north to south and from east to west.

treatment, thermal treatment, nonhazardous waste landfill, and hazardous waste landfill. Biological treatment is the cheapest disposal option, followed by thermal treatment and nonhazardous waste landfill options. Hazardous waste landfill is the most expensive and complicated disposal option.

Figure 5.41 is a solid 3D model of the soil sampling and excavation grid system. It is useful in visualizing the soil contamination situations at Site A and Site B, delineating the impacted soil boundaries, and calculating the volumes of the impacted soil categories. It can be rotated, tilted, and zoomed to allow readers to better visualize the grids from any angle, perspective, and distance. Furthermore, the individual grids inside the 3D model can be turned on or off to get a better view of other grids, especially those located inside the 3D model. For example, a group of grids in the southwestern corner of Site B (i.e., at the corner of Sherman Avenue and East 161st Street) were turned off so that the interior grids could be seen better without the grids blocking them. Also, a few top-level grids in the southeastern corner of Site A (i.e., at the intersection of Morris Avenue and East 161st Street) were turned off to allow better views of other grids blocked by them.

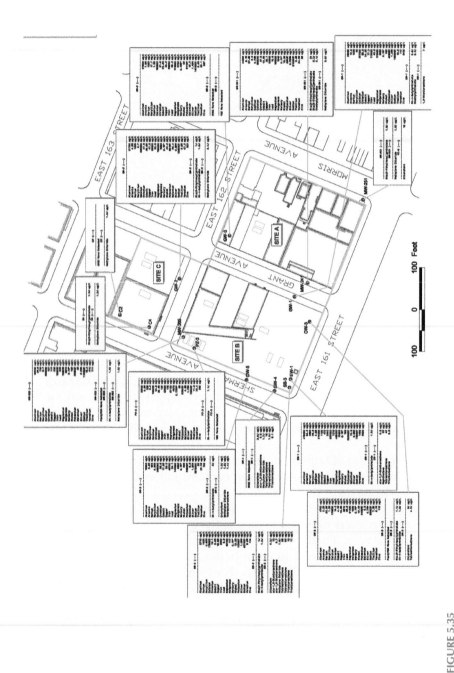

FIGURE 5.35

A data map showing the concentrations of VOC, SVOC, and metals in sampling event No. 3 groundwater samples in Site A, Site B, and Site C within the Bronx Criminal Court Complex project area in northern New York City. Samples in the southwestern area of the project sites, i.e., Site 2B and the western portion of Site 2A, contain higher concentrations of VOC, SVOC, and metals than samples in other project areas.

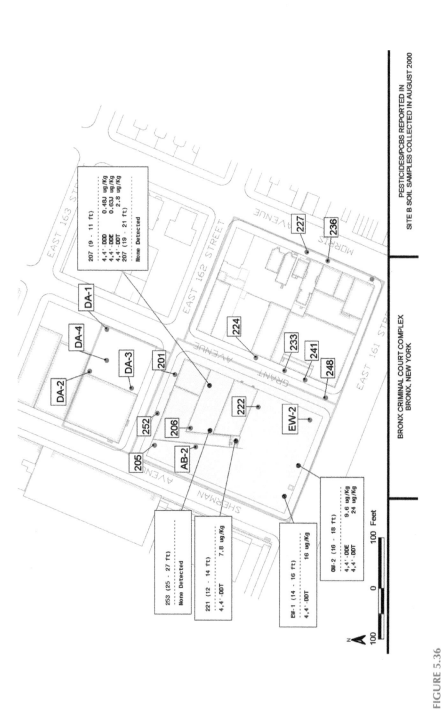

FIGURE 5.36

A data map showing the concentrations of pesticides and PCBs in soil samples collected in August 2000 from Site B of the Bronx Criminal Court Complex project area in northern New York City.

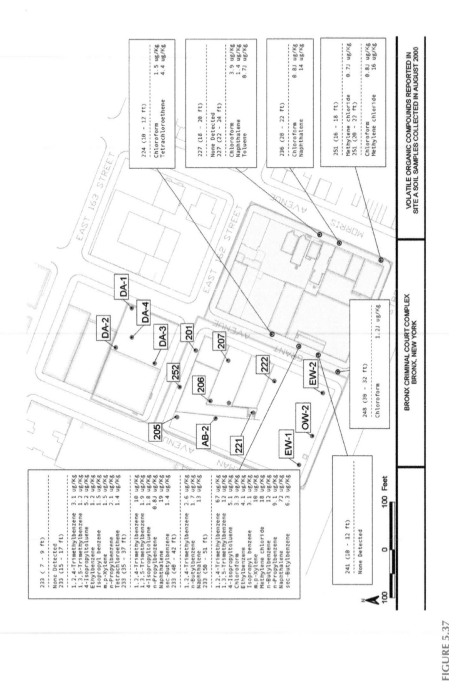

FIGURE 5.37

A data map showing the concentrations of volatile organic compounds (VOCs) in soil samples collected in August 2000 from Site A of the Bronx Criminal Court Complex project area in northern New York City.

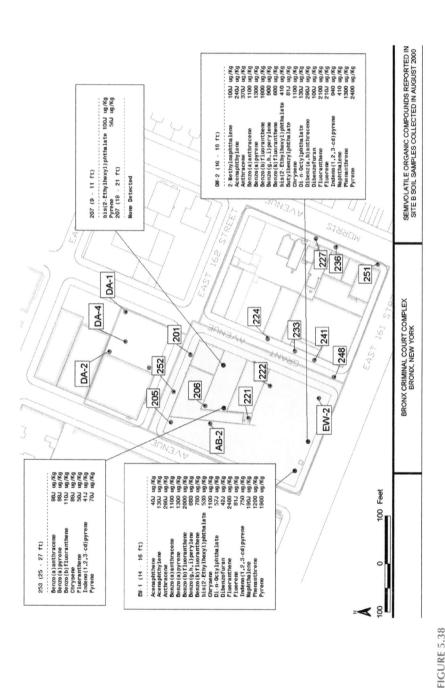

FIGURE 5.38

A data map showing the concentrations of semi-volatile organic compounds (SVOCs) in soil samples collected in August 2000 from Site B of the Bronx Criminal Court Complex project area in northern New York City.

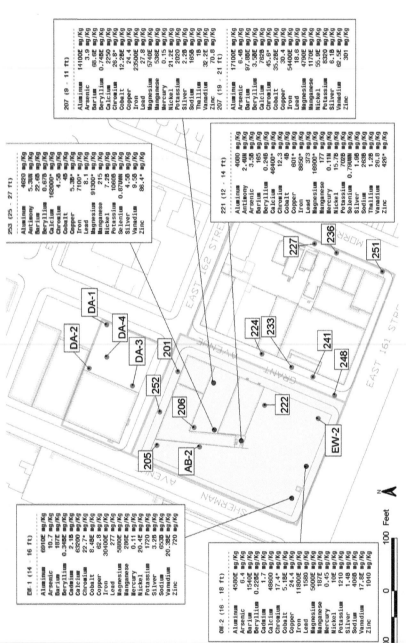

FIGURE 5.39

A data map showing the concentrations of metals in soils samples collected in August 2000 from Site B of the Bronx Criminal Court Complex project area in northern New York City.

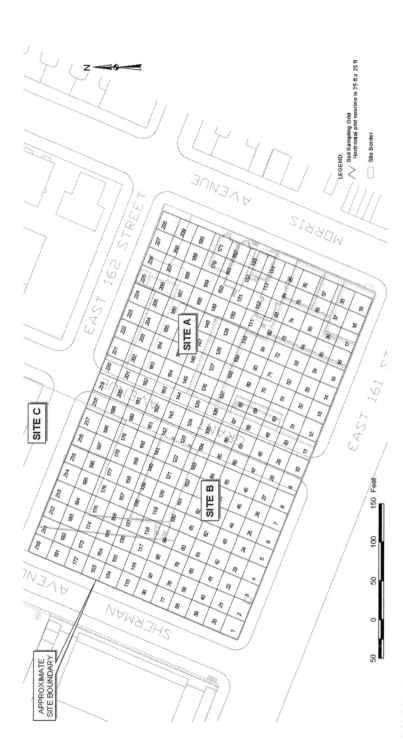

FIGURE 5.40
A sampling and excavation grid system in Site A and Site B of the Bronx Criminal Court Complex project area in northern New York City. Horizontal grid spacing is 25 feet × 25 feet. Samples were collected from three depth intervals, i.e., 0–5, 5–10, and 10–15 feet, respectively. The blue numbers are grid IDs.

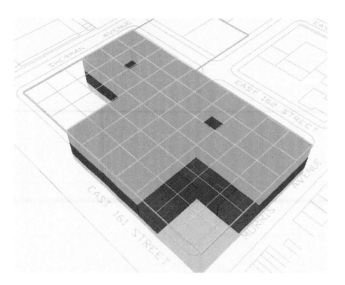

FIGURE 5.41

A solid 3D model of the soil sampling and excavation grid system in Site A and Site B of the Bronx Criminal Court Complex project area in northern New York City. Horizontal grid spacing is 25 feet × 25 feet. Samples were collected from three depth intervals: 0–5, 5–10, and 10–15 feet, respectively. Bright red represents hazardous oil. Light red represents petroleum-impacted soil. Blue represents nonhazardous soil (also called clean soil). Soil volume can be easily calculated from this 3D model. This 3D model can be rotated, tilted, and zoomed so that it can be visualized from any angle, perspective, and distance. Any 3D grids can be selectively turned on or off to better visualize other grids, especially those located inside the 3D model. For example, the grids in the southwestern corner of Site B (i.e., at the intersection of Sherman Avenue and East 161st Street) were turned off so that the interior grids could be viewed better. Also a few grids in the southeastern corner of Site A (i.e., at the intersection of Morris Avenue and East 161st Street) were turned off for the better views of other grids blocked by them.

Figure 5.42 is a hollow 3D model of the soil sampling and excavation grid system. With the interior soil not showing in this 3D hollow model, it is much easier to visualize soil contamination conditions in the whole 3D grid system. The colors on the walls of a grid represent the categories of the contaminations of the whole grid, that is, bright red representing hazardous oil, light red representing petroleum-impacted soil and blue representing nonhazardous soil. Both the hazardous oil and petroleum-impacted soil are called dirty soil, while the nonhazardous soil is called clean soil. Similar to the solid 3D grids model shown in Figure 5.41, the individual grids inside this 3D hollow model can also be turned on or off to have better view of other grids, especially those located inside the 3D model. In this 3D model, a group of grids in the southwestern corner of Site B were turned off to show the interior grids blocked by them.

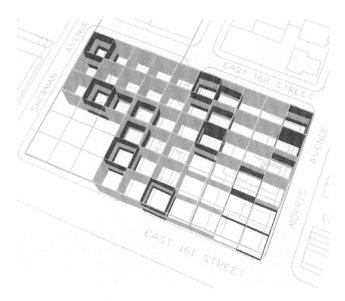

FIGURE 5.42
A hollow 3D model of the soil sampling and excavation grid system in Site A and Site B of the Bronx Criminal Court Complex project area in northern New York City. Horizontal grid spacing is 25 feet × 25 feet. Samples were collected from three depth intervals: 0–5, 5–10, and 10–15 feet, respectively. Bright red represents hazardous oil. Light red represents petroleum-impacted soil. Blue represents nonhazardous soil (also called clean soil). The colors of the walls of a grid represent the soil type of the whole grid at the specific depth interval, which was carved out for better visualization of the 3D grids.

Other Data Displaying and Visualizing Techniques

Although 2D and 3D analysis and modeling are important data visualization tools that have been explained in detail in this and other chapters, there are other useful data visualization techniques and approaches. Here, a few examples will be used to discuss these commonly used GIS techniques.

Using Tags, Pie Charts, and Color-Shaded Contours to Display Chemical and Geologic Data

In 2000, a small-scale environmental impact assessment investigation was conducted at a U.S. Coast Guard Buoy Depot in Massachusetts. Due to the high sensitivity of the facility, its name, location, and other related detailed information cannot be disclosed in this discussion. Sediment samples were collected and their analyzed results indicated that there were polycyclic aromatic hydrocarbons (PAHs) contaminants at the site. Since most PAHs are neither soluble in water nor volatile, they exist mainly in soil and sediments. Natural oil and coal deposits are major sources of PAHs. Depending on their structures, some are toxic while others are not toxic. There are about 20 chemicals in this category according to the U.S. Environmental Protection Agency (EPA). Therefore, each sediment sample was analyzed for many or all of these PAH chemicals. Due to their different toxicity properties, these PAH chemicals have a wide range of toxicity screening values (also known as thresholds). For these types of samples, one of the best ways to display and visualize the data is using tag maps, such as Figure 5.43.

A GIS program was written to extract the PAH chemicals and their analytical results from the master database for each sample and displayed them as a mini table (also known as a data tag) next to the sample location automatically. The chemicals exceeding the federal or state-issued official toxicity screening values were highlighted with red colors. The GIS program automated the tag map generating process, which would be very time consuming, tedious and prone to introducing errors if using traditional manual data reading and typing cartographic method. Since no human data reading and entry was involved in making the tag maps, no human error could be introduced in the mapping process. Therefore, only the master sampling database had to be checked to ensure data quality. It was confirmed that all the sample analytical data records stored in the master sampling database were correct and accurate through the strict data QA/QC procedure. With the GIS program, all the tag maps created with the data from the master database for the project should be correct and accurate, too. There was no need to review all the tag maps automatically generated, although a couple of tag maps were randomly selected for final QA/QC checking before all the maps were submitted to the federal and state regulators for their review and comment. The GIS program saved a significant amount of time in making and checking the data tag maps. The GIS program can be modified to process and display other types of data too.

With this type of data tag map, it is much easier to visualize and understand the overall contamination situations through the project site and examine the detailed information of individual samples. For example, with chemical analytical data displayed on the map, it is obvious to see which samples contain chemicals exceeding the federal- or state-issued official toxicity screening values due to their eye-catching bright red highlighting. It is not easy to miss them. Without this data processing and visualization technique, it would be difficult to understand the large volume of chemical analytical data records and relate them to the sampling locations on the project site. This data tag map indicates that PAH-related contamination was mainly in the soil and sediment in the south of the project site.

Metal contamination was discovered from samples collected from wetlands and swales too. The samples with metals exceeding the relevant federal- or state-issued toxicity screening values or thresholds were displayed as pie charts, as shown in Figure 5.44. The size of a pie chart is proportional to the total amount of the metals in the sample. The colors represent the

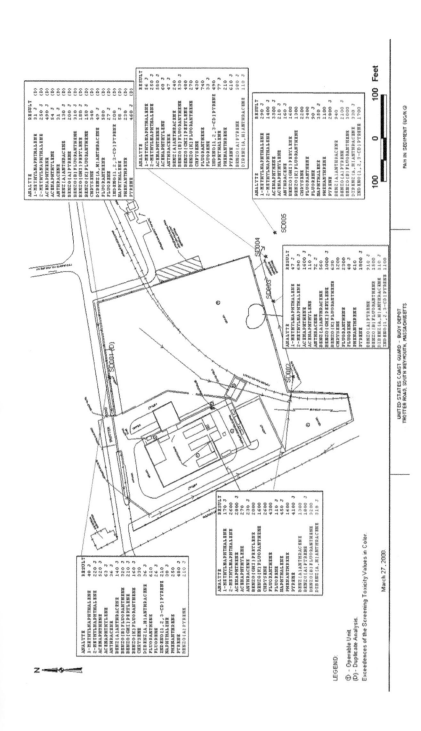

FIGURE 5.43

An example GIS tag map showing analytical results of PAHs chemicals in the sediment samples of a U.S. Coast Guard Buoy Depot in Massachusetts. The mini data tables (also commonly known as tags) contain the analytical results of the PAH chemicals of the sediment samples. The chemicals exceeding the toxicity screening values or thresholds were highlighted in red. The mini data tables and the highlighting were generated automatically from the master analytical database, with a GIS program written for this type of special format data labels.

metals: red for lead, green for copper, orange for arsenic, yellow for zinc, blue for nickel, and dark blue for chromium. Samples with no metal exceedance were shown as blue triangles.

With this kind of innovative GIS data processing and visualization technique, environmental scientists and regulators were able to quickly locate the samples with metals exceeding the relevant federal- or state-issued toxicity screening values, and see the amount of exceedance by looking at the colorful pie charts displayed on the map, instead of reading the large amount of chemical analytical data records and trying to relate them spatially to the sampling locations. By looking at this GIS map, it is obvious that lead and copper were the major contaminants in the wetlands and swales in this project site.

Many surface soil samples were collected to detect metal contamination. The analytical results of lead from the surface soil samples were statistically analyzed and displayed as a color-shaded concentration contours map, as shown in Figure 5.45. The Kriging (also known as Gaussian process regression) geostatistical interpolation method was mainly utilized to analyze the lead results in the surface soil samples. Several statistical analysis experiments were conducted, and each contour map was thoroughly reviewed and compared with the values in the metal analytical data tables to ensure that the statistical analysis results displayed on the map correctly and accurately represented the true contamination conditions of the sampling area.

The dark-red and bright-red shading represent higher lead concentrations in ppm (mg/kg), while the light-red shading represents lower lead concentrations in ppm. Although the statistically calculated shading contours covered the whole sampling area, the shading colors representing lead concentrations below 400 ppm were deliberately turned off (i.e., not shown) in the map for two good reasons. First, by doing so, this map emphasized the high lead concentration (i.e., $\geq$ 400 ppm) areas in the project site, which could exceed the relevant federal- and/or state-issued lead screening levels or thresholds and thus require remediation or further investigation. Second, by not showing the color shading in the very low lead concentrations (i.e., below 400 ppm) areas, this map tried to avoid possible misinterpretation that the whole sampled area had severe lead contamination issues, especially during public meetings.

Although data tag and pie chart maps are excellent methods of presenting analytical results of individual samples, a contour map, especially with eye-catching color shading, helps users to visualize the overall conditions of the whole project site by statistically interpolating the data from samples and spread the patterns and trends to cover the unknown (i.e., un-sampled) areas throughout the whole project site. With this type of color-shaded chemical concentration contour map, scientists and decision-makers can easily identify environmentally impacted areas (also known as hot spots), estimate their extents, and design mitigation and/or remedial actions.

Similar to environmental investigations, the color-shaded contouring technique is commonly used in geologic studies too. Figure 5.46 is a GIS map

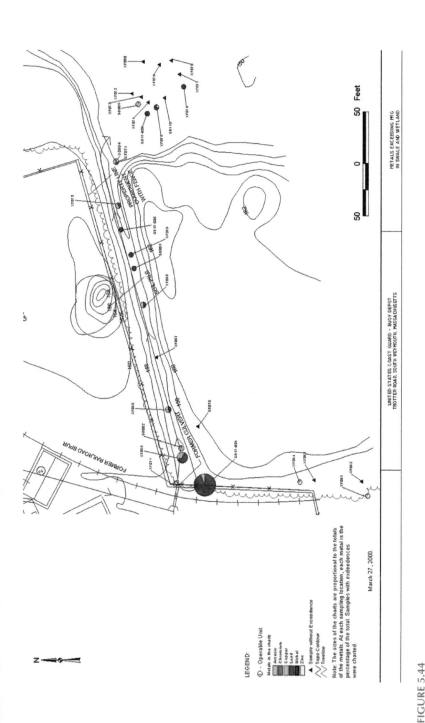

FIGURE 5.44

An example GIS map using pie charts to display analytical results of metals in samples collected from the wetlands and swales of a U.S. Coast Guard Buoy Depot in Massachusetts. The colors of the pie charts represent the metals exceeding the federal or state toxicity screening values or thresholds. The sizes of the pie charts are proportional to the total concentrations of the metals in the samples. Red represents lead, green for copper, orange for arsenic, yellow for zinc, blue for nickel, and dark blue for chromium. Blue triangles represent samples without any metal exceedance.

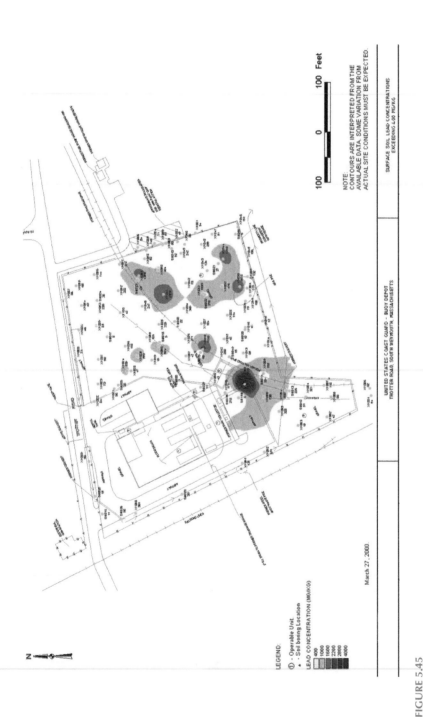

FIGURE 5.45

A color-shaded contour GIS map displaying the analytical results of lead in the samples collected from surface soil of a U.S. Coast Guard Buoy Depot in Massachusetts. The dark and bright red colors represent higher lead concentration in ppm (mg/kg) and the light red means lower lead concentration statistically interpolated from the analytical results of the samples. Although the statistically calculated color-shaded contours covered the whole project site sampled, the colors representing lead concentrations below 400 ppm were deliberated turned off (i.e., not showing) in order to emphasize the high lead concentration (i.e., ≥ 400 ppm) areas and avoid possible misinterpretation that the whole area had severe lead contaminations.

showing the top surface elevation of the bedrocks in Site 2A of the former Camp Sibert. As discussed earlier, Site 2A was used for chemical decontamination training in preparation for World War II. When the former Camp Sibert was closed in late 1945, chemical training materials, equipment, and unused chemical agents were buried at Site 2A. Previous investigations indicated that the soil was contaminated by the chemical training and burial activities. There was concern that the deep groundwater aquifer might be impacted, too. Therefore, a geologic study of the bedrocks of Site 2A was conducted in 2010. The elevation data of the top surface of the bedrocks from the deep water wells were statistically analyzed and displayed as color-shaded contours.

With these innovative data analysis and visualization approaches, such as data tag maps, pie chart maps, and color-shaded contour maps, it was much easier for environmental scientists, geologists, project managers, regulators, and other relevant stakeholders to visualize and understand the spatial patterns and trends of the geological and chemical conditions of the project sites and make the most appropriate mitigation and remediation decisions.

Linking Pictures, Water Well Diagrams, Geologic Cross Sections, Graphics and Other Information to Spatial Features in a GIS Map

GIS technology has the capability of linking external information, such as pictures, design drawings, data tables, text documents, and many others, to spatial features on a GIS map. These types of information help users to understand more about the features of interest. Figure 5.47 shows a wetland mapping project conducted in a New Jersey town in 2002. With the hotlink technique, field pictures were saved in the GIS database together with the wetlands GIS data files. When a wetland polygon is clicked with the hot-link tool, its picture shows up next to it. The pictures not only help users to understand the real-world conditions of the wetland area, they also support and confirm the correct and accurate interpretation and classification of the wetland types. Field classification errors could be corrected during the data QA/QC (quality assurance and quality control) process in office by more experienced and knowledgeable scientists with the assistance of the field wetland pictures. Without this type of data-linking technique, it would be difficult to save the large number of pictures together with their spatial GIS data. It would also be time-consuming to search hundreds or even thousands of pictures and find the right one needed to verify the wetland polygon whose classification was in doubt. It saves time and effort in storing both spatial and non spatial datasets, and in searching and displaying data.

For the Long-Term Monitoring and Sampling project in the Eastern Plume area inside the formerly used Naval Air Station Brunswick (NASB) in Maine, more than 40 groundwater monitoring wells were constructed as a part of the monitoring and sampling network to monitor the changing conditions within the sand groundwater aquifer which was impacted by dissolved-phase volatile organic compounds (VOCs). The wells were also used to pump

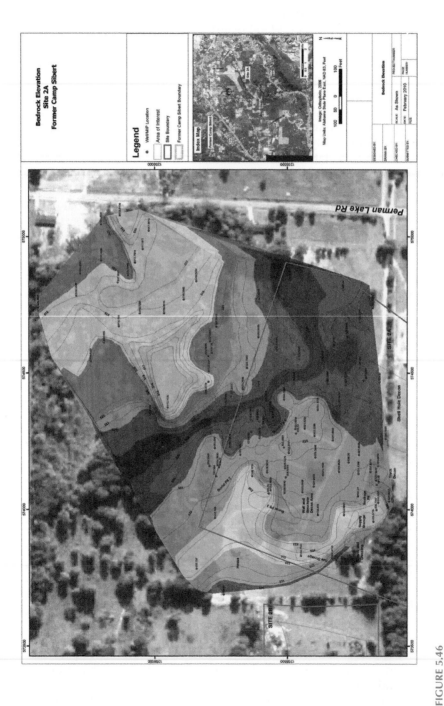

FIGURE 5.46

A color-shaded bedrock top surface elevation contour map of Site 2A inside the former Camp Sibert. The red dots are deep wells penetrating into the bedrocks and the numbers are bedrock top surface elevation above mean sea level (MSL), in feet. The colors represent the bedrock top surface elevation from the highest to lowest in the order of white, gray, brown, orange, yellow, green, and light-blue colors. The blue line is the boundary of Site 2A.

FIGURE 5.47

A GIS wetlands map developed during a wetland delineation project in a town in New Jersey. The cyan lines are the boundaries of the wetlands surveyed. The yellow points are GPS survey points. Field pictures were stored in the GIS database and linked to the wetland areas surveyed and mapped. When a wetland polygon is clicked with the hotlink tool, a picture of that wetland area appears next to the polygon. The linked pictures were useful in visualizing and understanding the wetland types and supporting the classification of the wetlands.

the contaminated groundwater by the treatment system, which operated from 1995–2000s. The groundwater contamination was caused by the past ordnance- and fuel-related activities in the formerly used NASB from 1943 to –2011, especially during the World War II and Cold War eras. Naval Air Station Brunswick was officially decommissioned in May 2011.

The detailed well construction diagrams of the more than 40 groundwater monitoring wells were scanned into Adobe Acrobat PDF documents and saved into the GIS database together with the water wells GIS data. The well construction diagram PDF files were linked to the water wells in GIS maps. When a well is clicked with the hotlink tool, its detailed well construction diagram shows up next to the well, as shown in the computer screenshot in Figure 5.48.

As discussed earlier, during the extensive geologic investigations in the former NASB, numerous geologic cross sections and fence diagrams were generated to study the geology of the project site. They were valuable datasets to help geologists, environmental scientists, project managers, regulators, and other stakeholders to understand the geologic and environmental conditions of the former NASB and make appropriate investigation and remediation decisions.

The useful geologic cross sections and fence diagrams were scanned into Adobe Acrobat PDF format files and saved into the GIS database. The scanned PDF files of the geologic cross sections and fence diagrams were linked to their cross section or fence diagram lines in GIS maps. When a cross-section or fence diagram line is clicked with the hotlink tool, its cross-section or fence diagrams show up in the GIS map, as shown in Figure 5.49. By displaying soil boring logs, water wells, the impacted area (i.e., the Easter Plume), hydrological, and other related features in a map, as well as geologic cross sections and/or fence diagrams, it is much easier to study the geologic conditions of the project site.

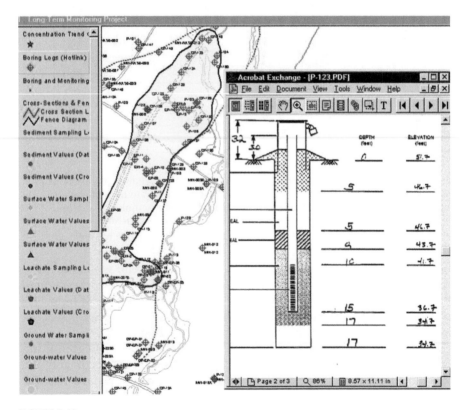

FIGURE 5.48
A groundwater wells GIS map created from the Long-Term Monitoring Project in the Eastern Plume area of the former NASB, Maine. Well construction diagrams were stored in the GIS database and linked to the water well locations on the map to help viewers to see the detailed information of the wells. When a well is clicked with the hotlink tool, its well construction diagram appears next to the well.

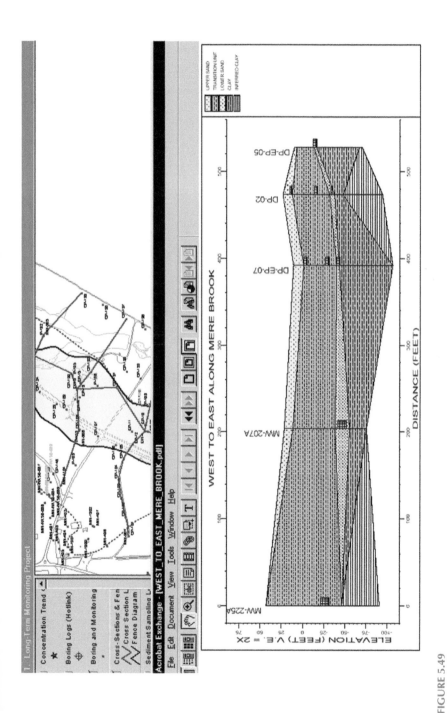

FIGURE 5.49

A GIS map showing geologic cross-section and fence diagram lines (in purple), created for the Long-Term Monitoring Project in the Eastern Plume area of the former NASB, Maine. Geologic cross sections and fence diagrams were stored in the GIS database together with soil boring logs and the water wells, and linked to the cross-section or fence diagram lines in GIS maps to help users to visualize the geologic conditions of the project site.

6

Sharing GIS Data and Maps through the Internet

The Internet is a global interconnected electronic communication network that uses Transmission Control Protocol/Internet Protocol (TCP/IP) to link billions of networks of organizations and individuals worldwide together via a broad array of wired and wireless networking technologies. It is a massive network on a global scale, consisting of smaller networks of millions of private, public, academic, business, and government networks. Through the Internet, various formats of information are exchanged and services are provided. These include online banking, publishing, ordering, reporting, dispatching, routing, tracking, advertising, video and audio streaming, gaming, social networking, cloud computing, and remote data storage, using the World Wide Web (WWW), electronic mail (e-mail), instant messaging, File Transfer Protocol (FTP), Voice over Internet Protocol (VoIP), and many other systems and applications.

Internet technology originated in the early 1960s from the research and development of the packet switching network, a digital networking communication method that groups transmitted data into suitably sized blocks, known as packets, to increase network efficiency. A few packet switching networks were developed. Among them was the Advanced Research Projects Agency Network (ARPANET) of the U.S. Department of Defense, which was developed in 1962 to facilitate internal communication via computer networks. In the 1980s, ARPANET made it possible to interconnect regional academic and military computer networks. With further improvement in networking technologies and worldwide participation in the 1990s, the modern Internet was born. Since then, it has been growing dramatically. Today, almost half of the world's population is using the Internet for a variety of purposes, and many organizations and individuals are joining the Internet and using it every day through computers, smartphones, smart tablets, smart TVs, gaming devices, and so on. With so many people and organizations using it, the Internet has become the most efficient and cost-effective way to distribute the most updated information and services.

Examples of Some Popular Geospatial Data and Mapping Websites and FTP Sites of the UN, EU, National and Local Government Agencies, Industries and Institutions

Many governmental agencies, organizations, industries, and institutions serve geospatial related datasets, maps, and services on the Internet, especially through WWW and FTP sites. From most public GIS websites and FTP sites, users can view, research, and create maps and download GIS datasets either totally free or by paying a fee. With some special GIS service websites or FTP sites, users can utilize the tools provided to download, upload, edit, and analyze GIS and related data to create customized maps, depending on their granted access permission level. Following is a list of a few popular and widely used geospatial data and mapping-related websites and FTP sites.

- **Maps and data website of the United Nations Operational Satellite Applications Program (UNOSAT)**

  The United Nations Operational Satellite Applications Program (UNOSAT) is a mapping and spatial data program under the United Nations Institute for Training and Research (UNITAR). UNOSAT was formed in 2003, mainly to support international humanitarian assistance operations and respond to crises caused by natural disasters, conflict, and other complex emergencies by supplying spatial data and maps whenever requested. UNOSAT is responsible for the rapid acquisition and processing of satellite imagery and data to generate specific information and analyses, create GIS data layers, and generate maps. Its rapid mapping and data services are available whenever they are needed, 24 hours a day, all year round. UNOSAT's ability to acquire satellite images and analyze them rapidly is crucial for humanitarian and disaster relief purposes.

  After relief operations are over, UNOSAT starts to support early recovery and development activities to help countries recover after disasters and crises. UNOSAT also offers GIS data and mapping assistance to some qualified national and international development projects. Some of UNOSAT's recent projects include supporting earthquake responses in Afghanistan and Pakistan, monitoring the Myanmar flooding, and improving disaster risk reduction in Bangladesh through capacity development in GIS and applications of satellite imagery.

  UNOSAT's GIS data and mapping website address (also known more formally as a uniform resource locator [URL]) is www.unitar. org/unosat/maps.

- **Data and maps website of the European Environment Agency (EEA)**

  The European Environment Agency (EEA) is an agency of the European Union that currently has 33 member countries. One of its primary tasks is to provide sound and independent information on the environment concerning the member countries. The EEA is a major information source for those involved in developing, adopting, implementing, and evaluating environmental policy, and also for the general public. The EEA's mandate includes helping the community and member countries make informed decisions about improving the environment, integrating environmental considerations into economic policies and moving toward sustainability, as well as coordinating the European environment information and observation network (Eionet).

  The EEA's data and mapping services are used by its 33 member countries, the European Commission, the European Parliament, the Council, the Economic and Social Committee, the Committee of the Regions, the business community, academia, nongovernmental organizations, other parts of civil society, and the general public. The EEA's data and maps website address is www.eea.europa.eu/data-and-maps.

- **The China Data Online website**

  Hosted by the China Data Center at the University of Michigan, the China Data Online website integrates China's historical, social, and natural science datasets into a Geographic Information System to facilitate comparative and interdisciplinary uses of the data; enable both nonspecialists and China scholars to understand and utilize these data; and enhance knowledge of China. The China Data Center was founded in 1997 as an international center for advancing the study and understanding of China. The missions of the China Data Center include supporting research in the human and natural components of local, regional, and global change; promoting quantitative research on China studies; facilitating collaborative research in spatial studies; and promoting the use and sharing of Chinese data in teaching and research. Collaborating with several Chinese government agencies and companies, the China Data Center distributes Chinese statistical data and publications and provides data services outside of China. The China Data Online website address is http://chinadataonline.org/.

- **India National Spatial Data Infrastructure (NSDI)**

  In 2000, the Department of Science and Technology established a task force under the chairmanship of the Surveyor General of India to create the National Spatial Data Infrastructure. It provides dual serials of maps, one for restricted uses by the armed forces and another for unrestricted civilian uses, with different georeferencing systems, projection systems, and sheet numbering systems. The web address of India NSDI is https://nsdiindia.gov.in/nsdi/nsdiportal/index.jsp.

- **Australian Government public datasets: Data.gov.au**

  Hosted by the Department of Finance, this website provides an easy way to locate, access, and reuse public datasets from the Australian government and state and territory governments. Its web address is https://data.gov.au.

- **Canada Lands Surveys Data website**

  This website is hosted by Natural Resources Canada. It contains a wide variety of data and maps of Canada. Its web address is www.nrcan.gc.ca/earth-sciences/geomatics/canada-lands-surveys/canada-lands-survey-system/10870.

- **Japan Maps and Geospatial Information website**

  This website is managed and hosted by the Geospatial Information Authority of Japan (GSI), a government agency of the Ministry of Land, Infrastructure, Transport and Tourism. It publishes spatial datasets and maps of Japan, such as the topographical 1:25,000 map series that covers the whole country, land use maps, land condition maps, volcanic land condition maps, maps of active faults in urban areas, aerial photographs, and so on. Its web address is http://www.gsi.go.jp/English.

- **United Kingdom spatial data: Data.gov.uk**

  Led by the Transparency Board, "data.gov.uk" is a key part of the United Kingdom government's work on transparency. This website brings together in one searchable website available datasets (including spatial data) from all UK central government departments and a number of other public sector bodies and local authorities. Its web address is https://data.gov.uk. Under this national data website, the Scottish Government created its own Spatial Data Infrastructure (Scottish SDI) that provides various maps and spatial datasets. Its web address is https://data.gov.uk/publisher/scottish-government-spatial-data-infrastructure.

- **National Geospatial Information (NGI) of the Republic of South Africa**

  National Geospatial Information (NGI), also known as South Africa's National Mapping Organization, is a component of the Department of Rural Development and Land Reform (DRDLR). It was created to establish an integrated survey system and perform extensive mapping coverage of the country. Its web address is www.ngi.gov.za.

- **New Zealand's catalogue of publicly funded geospatial data: geodata.govt.nz**

  "Geodata.govt.nz" is a core part of New Zealand's developing spatial data infrastructure. It was developed by the New Zealand Geospatial Office of the Ministry of Science and Innovation (MSI) and the National Institute of Water and Atmospheric Research (NIWA). It is web address is http://geodata.govt.nz.

- **Nigerian national Geo-Information website**

  Maps and geospatial data of Nigeria can be found from the website of the Office of the Surveyor General of the Federation (OSGOF), under the Federal Ministry of Works. Its web address is www.osgof.gov.ng.

- **Saudi Arabia Geological Survey maps and spatial data website**

  The Saudi Geological Survey (SGS) is an independent entity attached to the Ministry of Petroleum and Mineral Resources of the Kingdom of Saudi Arabia (KSA). The tasks and duties of the SGS include mapping, mineral exploration, geo-hazard and geo-environmental studies, hydrogeological studies, and so on. Digital datasets and maps are available from its website at www.sgs.org.sa/English/Pages/default.aspx.

- **Satellite images, maps, and data website of the NASA Earth Observatory**

  The Earth Observatory is an agency of the U.S. National Aeronautics and Space Administration (NASA). The Earth Observatory's mission is to share with the public the images, stories, and discoveries about the environment, Earth systems and climate that result from NASA research and explorations, including its satellite missions, field research, and models. The website of NASA satellite images is http://earthobservatory.nasa.gov/Images. Besides images, the website also hosts global-scale maps featuring a wide range of topics, such as land, water, atmosphere, climate, natural disasters, and so on.

- **Data.Gov: The home of the U.S. Government's open data**

  "Data.gov" website is managed and hosted by the U.S. General Services Administration's (GSA) Office of Citizen Services and Innovative Technologies. It is a public gateway to U.S. federal data holdings that are updated regularly. The Federal Open Data Policy requires that newly generated government data be made available in open, machine-readable formats while continuing to ensure privacy and security. Users can quickly and easily create maps and mashups from live web services and local content. Besides datasets, users can also find tools and resources to conduct research, develop web and mobile applications, design data visualizations, and so on. Its web address is www.data.gov.

- **USEPA's Environmental Dataset Gateway (EDG)**

  The United States Environmental Protection Agency's (USEPA) EDG is a gateway to web-based geospatial information and information services. It enables data consumers to discover, view, and access geospatial resources (e.g., data, services, or applications) made available by the EPA's program offices, regions, and labs. Its web address is www2.epa.gov/geospatial/epa-geospatial-data.

- **U.S. Census Bureau's Maps and Data website**

  This body provides the Census Data Mapper, a web mapping application intended to provide users with a simple interface to view, save, and print county-based demographic maps of the United States. Its web address is: http://datamapper.geo.census.gov/map.html. It also publishes the well-known and widely used TIGER Geodatabases, which are spatial extracts from the Census Bureau's Master Address File/Topologically Integrated Geographic Encoding and Referencing (MAF/TIGER) database for use with ESRI's ArcGIS. The 2015 TIGER Geodatabases can be searched and downloaded from U.S. Census Bureau's FTP site at ftp2.census.gov/geo/tiger/TGRGDB15. The U.S. Census Bureau also provides many other types of maps, data, and software for the general public, such as Reference Maps, Thematic Maps, Statistical Data Tools, Census Geocoder, Partnership Shapefiles, Relationship Files, Gazetteer Files, Block Assignment Files, Name Look-Up Tables, Tallies, LandView, and many others. Some data and maps are free, while others are available for a fee.

- **GIS Web Services for the FEMA National Flood Hazard Layer (NFHL)**

  The Federal Emergency Management Agency (FEMA) is an agency of the United States Department of Homeland Security. The National Flood Hazard Layer GIS service is available through FEMA's GeoPlatform, an ArcGIS Online portal containing a variety of FEMA-related data. Its web address is https://hazards.fema.gov/femaportal/wps/portal/NFHLWMS. The National Flood Hazard Layer is a computer database that contains FEMA's flood hazard datasets, depicting flood hazard information and supporting data used to develop the information. The NFHL data layers include flood hazard zones and labels; river miles markers; cross sections and coastal transects and their labels; letter of map revision (LOMR) boundaries and case numbers; Flood Insurance Rate Map (FIRM) boundaries, labels, and effective dates; coastal barrier resources system (CBRS) and otherwise protected area (OPA) units; community boundaries and names; levees; hydraulic and flood control structures; profile and coastal transect baselines; limit of moderate wave action (LiMWA); and other layers, too.

- **The USDA Geospatial Data Gateway (GDG)**

  The United States Department of Agriculture (USDA) Geospatial Data Gateway (GDG) provides environmental and natural resources datasets. The Gateway allows users to choose their area of interest, browse, select data from the catalog, customize the format, download selected datasets or have the datasets delivered on CDs or DVDs. Its web address is https://gdg.sc.egov.usda.gov. Their major data layers include aerial photos, topographic images, elevation, soil, easements, disaster events, climate, Public Land Survey System, geology, hydrography, geographic names, hydrological units, land use/land cover, government units, transportation, and so on. After the data are ordered through the website it processes the order and emails the user the FTP site address with data download instructions. Usually, downloading data through the FTP site(s) is free if the ordered data size is not too large. Otherwise, data on CDs or DVDs can be purchased.

- **The USGS Geo Data Portal (GDP)**

  The U.S. Geological Survey (USGS) Geo Data Portal (GDP) project provides scientists and environmental resource managers with access to downscaled climate projections and other data resources that are otherwise difficult to access and manipulate. Its web address is http://cida.usgs.gov/gdp. Users can also search and download a large variety of spatial-related data and maps from USGS EarthExplorer application. Its web address is http://earthexplorer.usgs.gov. The available dataset and maps depend on the area of interest. They can go as far back as the 1800s. USGS also distribute their latest version topographic maps in GeoPDF file format, historical aerial photos and topographic digital raster graphics (DRG), and other spatial datasets and maps through the Map Locator and Downloader website at http://store.usgs.gov.

- **Wetlands Mapper of the U.S. Fish and Wildlife Service**

  The U.S. Fish and Wildlife Service is a government agency of the U.S. Department of the Interior. Wetlands Mapper is a web-based application that integrates digital data with other resource information to produce wetlands-related management and decision support tools. It is a part of the National Wetlands Inventory program. Users can view wetlands and other related data, create and print maps, and download data layers. Its web address is www.fws.gov/wetlands/data/mapper.HTML.

- **US Army Corps of Engineers (USACE): Geospatial Portal**

  The USACE Geospatial Platform provides civil works wall maps, USACE ArcGIS Online Webmap (which enables map content creation, sharing, management, and manipulation), CorpsMap Viewer, which supports visualization and analysis of USACE infrastructure, and real time display of atmospheric, coastal, critical infrastructure, and watershed data. Its web address is http://geoplatform.usace.army.mil/home/.

- **The GeoCommunicator Land Survey Information System (LSIS)**

  The Department of the Interior's Bureau of Land Management (BLM) hosts and maintains a website for the distribution of Public Land Survey System (PLSS) data to support the mapping of federal land parcels. The Bureau of Land Management Cadastral Survey Program is responsible for the official boundary surveys of all federal-interest lands in the United States, which is over 700 million acres nationwide. The Public Land Survey System is the foundation for many survey-based Geographic Information Systems. Its web address is www.geocommunicator.gov/Geocomm/lsis_home/home.

- **NOAA Coast Survey GIS website**

  This GIS website is maintained and hosted by the U.S. National Oceanic and Atmospheric Administration (NOAA) Office of Coast Survey. It provides a variety of GIS maps, data, services, and tools. Its web address is www.nauticalcharts.noaa.gov/staff/gisintro.htm.

- **Google Maps website**

  Developed by Google, this is a very popular web-based mapping service, with roads covering the whole world. In many places, aerial or satellite imagery and street views (photos) are also available. Its web address is www.google.com/maps.

- **ESRI ArcGIS Online**

  ArcGIS Online is a commercial web-based GIS service product of the Environmental Systems Research Institute (ESRI). It allows subscribed users to integrate their data with ESRI's online maps, data layers and analytics to generate maps, publish data as web layers, collaborate and share GIS maps and data. The web address to sign up for a 60-day free trial is www.esri.com/software/arcgis/arcgisonline/evaluate.

- **New York State GIS Clearinghouse**

  This GIS website provides statewide New York GIS data and maps, created and maintained by various state and local governments. Its web address is https://gis.ny.gov.

- **Alabama Maps website**

  This website is hosted by the Cartographic Research Laboratory of the College of Arts and Sciences at The University of Alabama. It contains contemporary and historical maps, interactive maps, and aerial photos of the state of Alabama. Its web address is http://alabamamaps.ua.edu/index.html.

- **GIS Data and Maps: The Atlanta Regional Commission (ARC)**

  The Atlanta Regional Commission (ARC) is the regional planning and intergovernmental coordination agency for the 10-county area including Cherokee, Clayton, Cobb, DeKalb, Douglas, Fayette, Fulton, Gwinnett, Henry and Rockdale Counties, as well as the City of Atlanta in the state of Georgia. The ARC's GIS Data and Maps website is managed and hosted by the Research and Analytics Division in the Center for Livable Communities. It provides GIS data and maps to planning staff, local governments, planning partners, and the public. The web address is www.atlantaregional.com/info-center/gis-data-maps.

- **Fulton County (Georgia) GIS website**

  The Fulton County GIS website serves all departments of the county government and the public by providing a variety of online printable maps, interactive maps, and geospatial data, such as county-wide tax parcels, voting polls, structural assets, zoning, street centerlines, municipal boundaries, commission districts, land lots, traffic signals, library locations, county boundaries, topographic contours, comprehensive plan maps, water/sewer networks, and so on. Its web address is www.fultoncountyga.gov/fcgis-home.

- **Harvard University Center for Geographic Analysis (CGA)**

  Through this website, users can search and download a wide variety of GIS datasets from Harvard University and external sources. Its web address is www.gis.harvard.edu.

Case Study 6.1: A GIS Web-Based Application Developed for the Environmental Investigations at the Former Camp Sibert

As discussed earlier, during World War II (early 1942 to late 1945) former Camp Sibert (encompassing a land of 37,035 acres) was used for both conventional and chemical weapons training and other related activities. Due to the training activities and the burials of the chemical training materials, many areas (also known as sites) were contaminated by chemical warfare materials (CWM) and munitions and explosives of concern (MEC). There are around 40 chemical and conventional sites inside the former military installation. In the past 18 years, multiple environmental investigations and remediation actions have been conducted inside the former Camp Sibert. For such a large formerly used defense site (FUDS) with so many impacted areas/sites and environmental investigations, a wide range of stakeholders were involved, such as the owners (also known as caretakers) of the project sites; project contractors and their subcontractors; laboratories; suppliers; property owners of the impacted areas; businesses; the relevant federal, state, and local government agencies; and the general public. In order to communicate with the stakeholders efficiently and cost-effectively, a GIS web application was developed for the former Camp Sibert's environmental investigations and remediation actions. A computer screenshot of this GIS web application is shown in Figure 6.1. With a web browser, users could easily view and search the most updated datasets and print maps of the projects. For users without access to the Internet, this web application can be installed on their local computers or launched from CDs/DVDs or other storage media. A wide variety of GIS data layers and maps were uploaded into this web application and distributed to the users through the Internet, such as recent and historical aerial photos, topographic images, elevation and chemical concentration contours, roads, census data, water wells, wetlands, habitats of sensitive/protected species, parcels and right of entry (ROE), soil, geology, hydrology, groundwater elevation contours and flow directions, project sites/areas of interest, brush clearance, geophysical survey areas and transects, anomalies, intrusive investigation grids or trenches, samples, risk analysis maps, project status maps, investigations results maps and many more. Since this web application was integrated with the project master GIS database, any updates of the GIS data layers and maps were automatically reflected in the web-based datasets and maps so that all related stakeholders could see and use the most updated data and information to make real time, educated decisions.

In this example, the GIS web application as shown in Figure 6.1 has six GIS data layers, including a most recent aerial photo background, the county boundary (the light yellow line with a black dashed line on top of it), the former Camp Sibert boundary (the thick orange line), project sites (the blue polygons), historical features/areas of concern (small red polygons) and parcels (the black polygons). When a parcel is clicked, the detailed information about the property and its owner appears, such as the names and addresses of the owners, the parcel ID, the ROE status, and so on. When a project site polygon is clicked, detailed information on the site appears. If there is an active environmental investigation or remediation action on a site, detailed information about the ongoing project and its fieldwork status maps are available on the project GIS website for users to view and print, such as the latest version ROE status map (Figure 6.2), proposed and completed brush clearance status map (Figure 6.3), planned intrusive investigation and affected properties map (Figure 6.4), and comprehensive daily or weekly field work status map showing geophysical and intrusive investigation results, sampling and other activities (Figure 6.5).

In the ROE status map, as shown in Figure 6.2, the parcels shaded with dark green indicate that their ROE had been obtained, while the parcels shaded with light purple mean their ROE had not been granted yet. Efforts were under way to obtain their ROE. Therefore, all field work and related activities could only be performed in the properties with ROE and avoid those

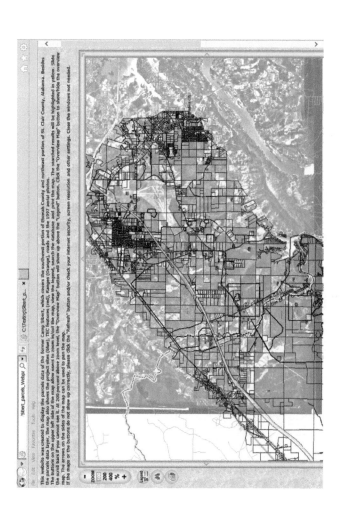

FIGURE 6.1

Screenshot of an example GIS web application developed for the environmental investigations inside the former Camp Sibert. It can be hosted on a web server, installed on local computers or launched from any storage media, such as external hard drives, CDs/DVDs, USB drives and so on. With a web browser, users can view and search the GIS data and print maps and other information. The map legend shows up when the Legend tool button is clicked. On this specific GIS web map, there are six data layers, i.e., the most recent aerial photo background, the former Camp Sibert boundary (thick orange line), project sites (blue polygons), features interpreted from historical aerial photos (small red polygons), parcels (black polygons) and the county's boundary (thick light yellow line with dashed black line on top of it). When a project site is clicked, detailed information of the site shows up. When a parcel is clicked, the detailed information about the property shows up, such as owners, addresses, parcel ID, right of entry (ROE) etc. Maps may also be available on the website if there is an active ongoing investigation or remediation project. Whenever a data layer is updated in the master GIS database, all related web maps and datasets are also updated automatically in real time. Therefore, all users view and use the most recent data and maps.

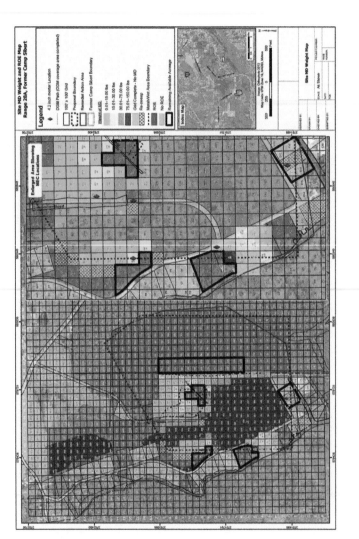

FIGURE 6.2

Right of entry (ROE) status map of a project site inside the former Camp Sibert, published on the GIS website. The inset map on the right side of the main map shows the enlarged area where 4.2 in. mortar rounds were found in the locations represented by the red diamond symbols. The large red polygon is the remedial action (also known as removal action) area inside the site. The dark green parcels have ROE, while the purple parcels do not have ROE. The brown area was surveyed by the Digital Geophysical Mapping (DGM). The little squares are 100 × 100 feet grids with their IDs (white) inside them. The colored squares are the grids which have been intrusively investigated so far. Their colors represent the amount of munitions debris (MD) in pounds (lb) dug out from the grids, ranging from 150 to 0.01 lb, in the descending order of red, orange, yellow and green. The cyan grids were completely investigated, with no MD found in them. The black polygons are areas pending geophysical and/or intrusive investigations.

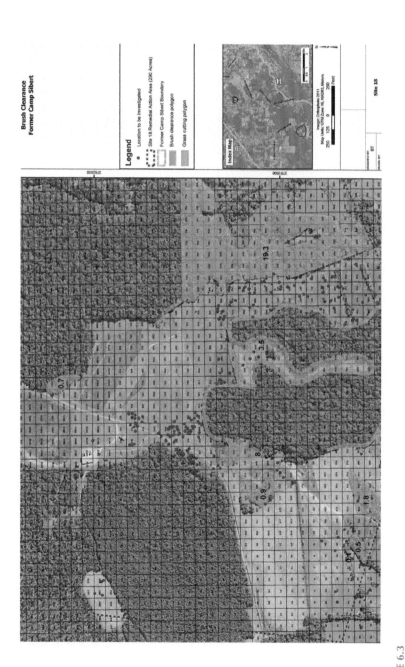

FIGURE 6.3

Brush clearance status map of a project site inside the former Camp Sibert, published on the GIS website. The red dots are selected geophysical anomaly locations to be intrusively investigated. The dashed red line polygon is the remedial action area inside the project site. Purple designates the area needing brush clearance, while green represents the area requiring grass cutting. The numerical labels are acreage numbers of the brush clearance or grass cutting areas.

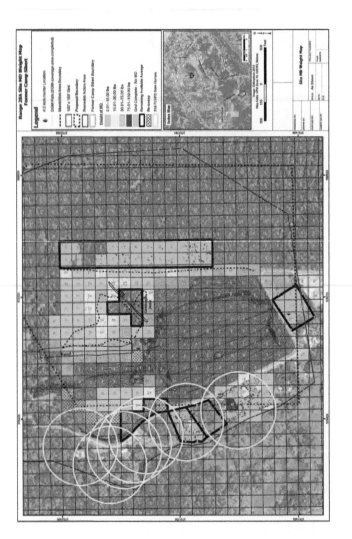

FIGURE 6.4

A GIS map showing the safety zones of a project site inside the former Camp Sibert, published on the GIS website. The large red polygon is the remedial action (also known as removal action) area inside the site. The yellow circles represent safety zones (also known as safety buffers, or minimum separation distance zones) around the affected homes (also known as occupied structures). Its distance or radius depends on the type of the munitions and explosives of concern (MEC). In this specific site, it is 316 feet for the 4.2 in mortar rounds. Remedial action has to be performed outside the safety zones. Otherwise, approved engineering control measures must be used, or the buildings have to be evacuated. The brown area was surveyed by the Digital Geophysical Mapping (DGM). The little squares are 100 feet × 100 feet grids with their IDs (white) inside them. The colored squares are the grids which were intrusively investigated. Their colors represent the total amount of munitions debris (MD) in pounds (lb) dug out from the grids, ranging from 150 to 0.01 lb, in the descending order of red, orange, yellow and green. The cyan grids were completely investigated, with no MD found in them. The black polygons are areas pending geophysical investigations.

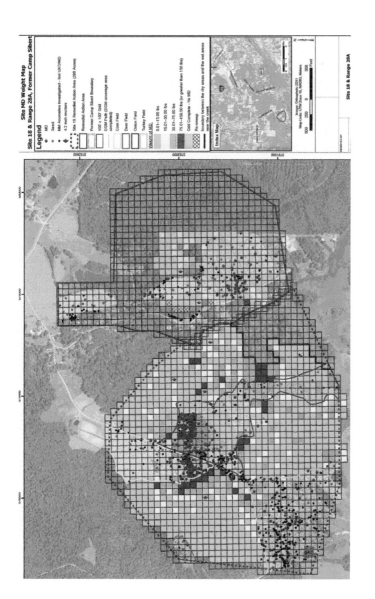

FIGURE 6.5

A comprehensive GIS map displaying the field work status of two remedial action project sites inside the former Camp Sibert, published on the GIS website. The colored dots are anomalies that were identified by the digital geophysical mapping (DGM) method and intrusively investigated. The small squares are the 100 × 100 feet grids to be either intrusively investigated or geophysically surveyed. The colored squares are the grids which had been intrusively investigated so far. Their colors represent the total amount of munitions debris (MD) in pounds (lb) dug out from the grids, ranging from 150 to 0.01 lb, in the descending order of red, orange, yellow and green. The cyan grids were completely investigated, but with no MD items found in them. This status map was updated on daily or weekly basis depending on the field work schedule, weather conditions and/or other factors.

properties without ROE. Whenever the ROE status changes, the web map is updated automatically in real time. All the parties involved, especially the field workers, were able to see exactly the same most recent information published on the website. It saved money and time by distributing and communicating information through this GIS website instead of notifying the related stakeholders individually by mails, phone calls, fax, or other means.

Besides ROE status information, this map also shows other data layers, including the locations of the 4.2 in. mortar rounds discovered and retrieved (the red diamond symbols), digital geophysical mapping (DGM) paths and areas, 100 feet × 100 feet intrusive investigation grids (i.e., the 100' × 100' Grid in the legend), boundary of the munitions and explosives of concern (MEC) remedial action area (the red line), the Camp Sibert boundary (thick gold line in the Index Map), results (also known as findings) of intrusive investigation in the grids represented with different colors based on the weight of munitions debris (MD) recovered from the grids, boundary of wet and flooded areas, and so on. Since the project site area with the 4.2 in. mortar rounds discovered was of special importance, an inset map was created to show an enlarged view of it so that users can see more details of the area.

For a project site with thick vegetation coverage in the planned investigation and/or remedial action areas, brush clearance has to be performed. Figure 6.3 is an example brush clearance map published on the former Camp Sibert GIS website. The thick dashed gold line in the Index Map represents the boundary of the former Camp Sibert. The red dashed line on the main map is the boundary of the remedial action area of the project site. The red points represent the locations of the geophysical anomalies which were selected for intrusive investigation. The area shaded with light green required grass-cutting, while those areas shaded with purple needed brush clearance. The numbers on the grass-cutting and brush-clearance areas are their acreage numbers. This brush-clearance status map was updated on either a daily or weekly basis depending on the project schedule, weather conditions and other factors. All parties involved in the project and field work—such as the project manager and staff, field crews, brush and grass-cutting subcontractors and owners of the affected properties—were able to see and use this updated map for information and to help perform the relevant tasks.

In the planned intrusive investigation and affected properties map (as shown in Figure 6.4), the yellow circles are safety zones around the affected homes, also known as occupied structures. The circle radius (also known as safety zone distance) depends on the nature of the MEC hazards documented, reported, or suspected on the site. There are official guidelines for calculating the distances of safety zones, also known as safety buffers or minimum separation distance (MSD). In this specific site, the radius of the safety zones around the affected homes was calculated as 316 feet based on the relevant guidelines. Following the guidelines, intrusive investigations have to be conducted outside the safety zones around the affected homes. Otherwise, approved engineering control measures have to be taken to protect the homes, or the occupants have to be evacuated for the duration of time the work takes.

In addition to the affected properties and the safety zones around the affected homes, this map also shows other important data layers. The red diamond symbols in the map represent locations of the 4.2 in. mortar rounds discovered and retrieved at the site. The densely spaced brown lines represent digital geophysical mapping (DGM) paths. The dark red lines are 100 feet × 100 feet intrusive investigation grids (i.e., the 100' × 100' grid in the legend). The intrusive investigation results in the grids were displayed with different colors based on the total weight of munitions debris (MD) recovered from the grids. The red line is the boundary of the MEC remedial action area. The dashed magenta line represents the proposed MEC remedial action area. The dashed black line marks the boundary of wet and flooding areas. The areas inside the thick black lines needed geophysical investigation. The thick gold dashed line in the Index Map is the former Camp Sibert boundary.

Figure 6.5 is an example comprehensive field work status map of two remedial action project sites within the former Camp Sibert. It displays the most updated geophysical survey and intrusive investigation results. The color points represent geophysical anomalies which were intrusively investigated and confirmed. For example, the red points indicate anomaly locations where munitions debris (MD) items were found. Blue points represent seeds (also known as blind seeds or QC seeds) uncovered. Seeds are items of inert munition or munitions surrogate placed on a project site to check the quality of a geophysical survey. The locations and identities of seeds are unknown to the people who perform the geophysical survey, data analysis and intrusive investigation. The black points are metal mapper anomalies that were intrusively investigated, but no unexploded ordnance (UXO) or MD were found. The red diamonds represent the 4.2 in. mortar rounds recovered. The colored grids illustrate the intrusive investigation results in the grids. The colors represent the total weight of munitions debris dug out from the grids. The solid and dashed red lines represent the boundaries of the two different MEC remedial action areas. The dashed magenta line represents the proposed MEC remedial action areas inside the former Camp Sibert.

Commonly Used Acronyms in GIS and Environmental Science

AB: air base
ABP: agent breakdown product
ACA: American Cartographic Association
ADDA: American Design Drafting Association
ADE: AutoCAD data extension
ADEO: advanced earth observing satellite
ADR: automated data review
A/E: architectural/engineering
A&E: architect and engineering
AGS: American Geographical Society
AI: artificial intelligence
AK: Alaska
AL: Alabama
AM: automated mapping
AOI: area of interest
APP: accident prevention plan
ArcFM: arc facilities manager
ASCII: American Standard code for Information Interchange
ASR: archives search report
AZ: Arizona
BG: block group
BLDG: building
BLM: U.S. Bureau of Land Management
BM: benchmark
.BMP: bitmap image format
B/W: black and white
CA: California; chemical agent
CACM: chemical agent contaminated media
CADD: computer-aided design and drafting
CAIS: chemical agent identification set(s)
CD: compact disc
.CDR: CorelDRAW format
CD-R: compact disk–recordable
CD-ROM: compact disk–read only memory
CERCLA: comprehensive environmental response; Compensation and Liability Act
CIA: U.S. Central Intelligence Agency
CIH: Certified Industrial Hygienist

CK: cyanogen chloride
COE: Corps of Engineers
CP: command post
CRREL: Cold Regions Research Engineering Laboratory
CSM: conceptual site model
CSS: chemical safety submission
CST: Central Standard Time
CWM: chemical warfare materiel
DAAMS: depot area agent monitoring system
DASAF: Department of the Army Safety Office
DD: data dictionary
DDESB: Department of Defense Explosives Safety Board
Diam.:   diameter
DIDs: data item descriptions
DLG: digital line graph
DNR: Department of Natural Resources
DoD: Department of Defense
DOQ: digital orthophoto quadrangle
DOQQ: digital orthophoto quarter quadrangle
DOT: Department of Transportation
DQCRs: daily quality control reports
DQOs: data quality objectives
DRG: digital raster graphic
DTD: digital terrain data
.DWG: native file format for AutoCAD
.DXF: drawing interchange file
EA: environmental assessment
Ed.:   editor; edition
EDDs: electronic data deliverables
EMTs: Emergency Medical Technicians
EPA: U.S. Environmental Protection Agency
EPP :   environmental protection plan
ER: engineer regulation
ERDAS: Earth resources data analysis system
ESRI: Environmental Systems Research Institute, Inc.
EZ :   exclusion zone
F: Fahrenheit
FAA: U.S. Federal Aviation Administration
FFP: firm fixed price
FHA: Federal Highway Administration (U.S.)
FIP: field investigation plan
FL: Florida
FTP: file transfer protocol
FUDS: formerly used defense sites
GA: Georgia

GFCI :	ground-fault circuit interrupter
GFE:	government-furnished equipment
.GIF:	graphics interchange file
GIS:	geographical information system
GMT:	Greenwich Mean Time
GNIS:	Geographic Names Information System
GNSS:	global navigation satellite system
GO/CO:	government owned/contractor operated
GPS :	global positioning system
GRASS:	Geographic Resource Analysis Support System
GSA:	General Services Administration (U.S.); Geological Society of America
GUI:	graphical user interface
GW:	groundwater
HD:	high definition
HDDT:	high-density digital tape
HDT:	high-density tape
HDTR:	high-density tape recorder
HDTV:	high-definition television
HF:	high frequency
HFA:	Hydrologic Field Assistant
.HGL:	Hewlett-Packard graphics language format
HGML:	hyper graphic markup language
HI:	Hawaii
HP:	Hewlett-Packard (corp.)
HQ:	headquarters
HRS:	Hazard Ranking System
HTML:	Hypertext (or HyperText) Markup Language
HTRW:	hazardous, toxic, and radiological waste
HUA:	hydrologic unit area
HUC:	hydrologic unit code
HUD:	Department of Housing and Urban Development (U.S.)
H/W:	hardware
IBM:	International Business Machines (Corp.); international boundary marker
IC:	integrated circuit(s)
ICA:	International Cartographic Association
ID:	identification; interface device; Idaho; island
IDIQ:	indefinite delivery/indefinite quantity
IDW :	investigative-derived waste
IIS:	Internet information server
IL:	Illinois
IM:	information management
.IMG:	ADRG image format (USDMA)
in.:	inch

IN:	Indiana
INLT:	inlet
INPR:	inventory project report
Inst.:	institute
I/O:	input/output
IP:	Internet protocol
IS:	International Standard
ISO:	International Standards Organization
ISP:	Internet service provider
ISPRS:	International Society for Photogrammetry and Remote Sensing
ISR:	information storage and retrieval
ISU:	International System of Units
JERS:	Japanese Earth Resources Satellite
JFIF:	JPEG file interchange format
.jgw:	JPEG with World File
JICA:	Japan International Cooperation Agency
JIGI:	Java Interface for Geospatial Information
JPEG:	Joint Photographic Experts Group
.JPG:	Joint Photographic Experts Group format
K:	kilo
kHz.:	kilohertz
km:	kilometer
kV:	kilovolt
kW:	kilowatt
KY:	Kentucky
m:	meter
MAPSAT:	mapping satellite (U.S.)
MB:	megabyte
MC:	munitions constituent
MCE :	maximum credible event
MCL :	maximum contaminant level
MD:	Maryland
MDA:	monochrome display adapter
MDS:	metadata standards
ME:	Maine
MEC:	munitions and explosives of concern
MGFD:	munitions with the greatest fragmentation distance
MHz:	megahertz
Mi.:	mile(s)
MI:	Michigan
Mil.:	Military
MIME:	Multipurpose Internet Mail Extension
MIS:	management information system
MM CX:	Military Munitions Center of Expertise

MMRP:	Military Munitions Response Program
MN:	magnetic north; Minnesota
MO:	Missouri
MPM :	most probable munitions
MRSH:	marsh
MRSPP:	Munitions Response Site Prioritization Protocol
MSD :	minimum separation distance
MS-DOS:	Microsoft disc operating system
MSL:	mean sea level
MST:	Mountain Standard Time
MT:	Montana
Mtn.:	Mountain
N.:	North
NA:	National Archives
NAD:	North American Datum
NAD27:	North American Datum of 1927
NAD83:	North American Datum of 1983
NADCON:	North American Datum Conversion
NAPA:	National Academy of Public Administration
NAPP:	National Aerial Photography Program (U.S.)
NARA:	National Archives and Records Administration (U.S.)
NARS:	National Archives and Records Service (U.S.)
NARSA:	National Advanced Remote Sensing Applications Program
NAS:	National Academy of Sciences (U.S.)
NASA:	National Aeronautics and Space Administration (U.S.)
NATO:	North Atlantic Treaty Organization
NAVD:	North American Vertical Datum
NAVD88:	North American Vertical Datum 1988
NC:	North Carolina
NCDCDS:	National Committee for Digital Cartographic Standards (U.S.)
NCDS:	National Cartographic Data Standard
NCDSB:	National Cartographic Data Standard Base
NCP:	National Contingency Plan
ND:	North Dakota
NDAI:	no Department of Defense action indicated
NE:	Northeast; Nebraska
NELAP:	National Environmental Laboratory Accreditation Program
NGRS:	National Geodetic Reference System (U.S.)
NGS:	National Geographic Society (U.S.)
NGSP:	National Geodetic Satellite Program
NGVD:	National Geodetic Vertical Datum
NGVD29:	National Geodetic Vertical Datum of 1929
NHD:	National Hydrography Dataset (USGS)
NHS:	National Highway System
NJ:	New Jersey

NM:	New Mexico
NMAS:	National Map Accuracy Standard(s)
NMP:	National Mapping Program (USGS)
NNE:	North northeast
NNSS:	Navy Navigational Satellite System
NNW:	north–northwest
NOAA:	National Oceanic and Atmospheric Administration (U.S.)
NOSE:	no significant effects distance
NOSS:	National Oceanographic Satellite System (U.S.)
NPS:	National Park Service (U.S.)
NRA:	National Recreation Area
NRC:	National Research Council
NRCS:	Natural Resources Conservation Service (USDA)
NRI:	National Resources Inventory (SCS)
NRS:	National Reference System (U.S.)
NSA:	National Security Agency
NSF:	National Science Foundation (U.S.)
NSN:	National Satellite Network
NSSDA:	National Standard for Spatial Database Accuracy
NSTL:	National Space Technology Laboratory
NTAD:	National Transportation Atlas Data (U.S.)
NTCRA:	non-time-critical removal action
NTP :	notice-to-proceed
NTSB:	National Transportation Safety Board (U.S.)
NV:	Nevada
NVCN:	National Vertical Control Network (U.S.)
NWI:	National Wetlands Inventory (USFWS)
NWS:	National Weather Service (NOAA)
OB/OD:	open burn/open detonation
ODE:	open development environment
OH:	Ohio
OK:	Oklahoma
OLE:	object linking/embedding
OR:	Oregon
OQ:	orthophoto quadrangle
OSHA :	Occupational Safety and Health Administration
PA:	Pennsylvania; preliminary assessment
PC:	personal computer
PDS :	personnel decontamination station
P.E.:	Professional Engineer
PEN:	peninsula
Penn.:	Pennsylvania
Pixel:	picture element
PK:	peak
PLS:	Professional Land Surveyor

PM:	project manager
PND:	pond
.PNG:	portable network graphics (format)
POC:	point of contact; point of control
PPE:	personal protective equipment
PQCM:	project quality control manager
PQCP:	Parsons Quality Control Plan
PRGs :	preliminary remediation goals
PSAP:	programmatic sampling and analysis plan
PSHM:	project safety and health manager
PSR:	project status report
PST:	Pacific Standard Time
PSU:	primary sampling unit
PT:	point
PWP:	programmatic work plan
PWS:	performance work statement
QA:	quality assurance
QA/QC:	quality assurance/quality control
QC:	quality control
QR:	qualitative reconnaissance
QSM:	quality systems manual
RADAR:	radio detecting and ranging
RAM:	random access memory
RBCs :	risk-based concentrations
RCRA:	Resource Conservation and Recovery Act
R&D:	research and development
RDBMS:	relational database management system
RDGE:	ridge
Rev:	revised
RGB:	red-green-blue
RI:	Rhode Island
RI/FS:	remedial investigation and feasibility study
RIP:	raster image processing
RMT:	remotely measured target
ROE:	right of entry
ROW:	right of way
RP:	reference point
RR:	rail road
R/S:	remote sensing
RSRT:	resort
RSV:	reservoir
RTF:	Rich Text Format
RT-GPS:	real-time differential GPS
RTK :	real-time kinematic
RTP:	real-time positioning

RUIN:	ruins
SBAS :	space base augmentation system
SC:	South Carolina
SCH:	school
SD:	South Dakota
SDB:	spatial data base
SDBMS:	spatial database management system
SDE:	Spatial Database Engine (ESRI)
SDES:	Spatial Data Exchange Standard
SDSFIE:	Spatial Data Standards for Facilities, Infrastructure, and Environment
SDTS:	Spatial Data Transfer Standard
SE:	Southeast
SEDD:	staged electronic data deliverable
SHOL:	shoal
SI:	site inspections
SLD29:	Sea Level Datum of 1929
SLF:	Standard Linear Format
SMTP:	Simple Mail Transfer Protocol
SOW:	statement of work
SPCS:	state plane coordinate system
SQL:	Standard Query Language; Structured Query Language
SSE:	south–southeast
SSF:	Trimble Standard Storage Format
SS-FSP:	site-specific field sampling plan
SSHO:	site safety and health officer
SS-QAPP:	site-specific quality assurance project plan
SSW:	south–southwest
SS-WP:	Site-specific work plan
St:	street
STDM:	stadium
STEL:	short-term exposure limit
STL:	Severn Trent Laboratory
SU:	sampling unit
SUV :	sport utility vehicle
SUXOS:	senior unexploded ordnance supervisor
TAR:	tape address register
TBD:	to be determined
TCLP :	toxicity characteristic leaching procedure
TCP/IP:	Transmission Control Protocol/Internet Protocol
TCRA:	time-critical removal action
TEC:	topographic engineering center
TESC:	tile enclosed elevator with self-closing traps
.TIF:	Tagged Image File
TIFF:	Tagged Image File Format

TIGER:	Topologically Integrated Geographic Encoding and Referencing (U.S. Census Bureau)
TIN:	triangulated irregular network
TMP:	technical management plan
TPP:	technical project planning
TN:	Tennessee
TRS:	Township/range system; Trimble Teference Station
TVA:	Tennessee Valley Authority
TWA:	time-weighted average
Twp.:	township
TX:	Texas
UA:	urbanized area
UK :	United Kingdom
UML:	unified modeling language
UN:	United Nations
UNDP:	United Nations Development Program
UNEP:	United Nations Environment Program
UNESCO:	United Nations Educational, Scientific and Cultural Organization
UNGEGN:	United Nations Group of Experts on Geographical Names
UNGIWG:	United Nations Geographical Information Working Group
UPC:	universal product code
UPPS:	universal projection plotting system
UPRN:	unique property reference number
URL:	universal resource locator
U.S.:	United States
USA:	United States of America
USACE:	U.S. Army Corps of Engineers
USACERL:	U.S. Army Construction Engineering Research Laboratories
USAESCH:	U.S. Army Engineering and Support Center, Huntsville
USAETL:	U.S. Army Engineer Topographic Laboratories
USAF:	U.S. Air Force
USATCES:	U.S. Army Technical Center for Explosives Safety
USBC:	U.S. Bureau of the Census
USC:	unified soil classification
USCE:	U.S. Army Corps of Engineers
USCG:	U.S. Coast Guard
USC&GS:	U.S. Coast and Geodetic Survey
USCS:	U.S. Customary System
USDA:	U.S. Department of Agriculture
USDC:	U.S. Department of Commerce
USDI:	U.S. Department of the Interior
USEPA:	U.S. Environmental Protection Agency
USDMA:	U.S. Defense Mapping Agency
USFS:	U.S. Forest Service
USFWS:	U.S. Fish and Wildlife Service

USGS:	U.S. Geological Survey
USNBS:	U.S. National Biological Survey
USN:	U.S. Navy
USPLSS:	U.S. Public Land Survey System
USPS:	U.S. Postal Service
USSCS:	U.S. Soil Conservation Service
UST:	underground storage tank
UT:	Utah
UTM:	Universal Transverse Mercator
UV:	ultraviolet
U-2:	high-altitude remote sensing aircraft
UXO:	unexploded ordnance
UXOQCS:	unexploded ordnance quality control specialist
VA:	Virginia
VBA:	Visual Basic for applications
VGA:	video graphics array; video graphics adaptor
VI:	Virgin Islands
VLC:	volcano
VML:	Vector Markup Language
VOC:	volatile organic compound
VT:	Vermont
WA:	Washington
WAAS:	wide area augmentation system
WGS:	World Geodetic System
WGS84:	World Geodetic System of 1984
WI:	Wisconsin
WMD:	water management district
.WMF:	Windows Metafile Format
WV:	West Virginia
WWW:	World Wide Web
WY:	Wyoming
XML:	Extensible Markup Language

Bibliography

Banerjee, S., Carlin, B.P., and Gelfand, A.E., *Hierarchical Modeling and Analysis for Spatial Data*, Boca Raton, FL: Chapman and Hall/CRC Press, Taylor and Francis Group, 2004.

Bardinet, C. and Royer, J.J., *Geoscience and Water Resources: Environmental Data Modeling*, Berlin: Springer, 1997.

Burdea, G. and Coiffet, P., *Virtual Reality Technology*, New York: Wiley, 1994.

Cruz-Neira, C., Leigh, J., Papka, M., and Barnes, C. et al., Scientists in wonderland: A report on scientific visualization applications in the CAVE virtual reality environment, *IEEE Symposium in Research Frontiers in Virtual Reality*, Visualization 93, San Jose, CA, 1993.

Douglas, W.J., *Environmental GIS Applications to Industrial Facilities*, New York: Lewis, 1995.

ESRI, *Working with the Geodatabase: Powerful Multiuser Editing and Sophisticated Data Integrity*, Redlands, CA: ESRI Press, 2002.

Harder, C., *Serving Maps on the Internet*, Redlands, CA: ESRI Press, 1998.

Haxel, G.B. and Dillon, J.T., The Pelona-Orocopia Schist and Vincent-Chocolate Mountain thrust system, southern California, in D.G. Howell and K.A. McDougall (eds.), *Mesozoic Paleogeography of the Western United States*, pp. 453–469, Los Angeles, CA: Society of Economic Paleontology and Mineral Pacific Sector, 1978.

Heipke, C.D. and Straub, M., Towards the automatic GIS update of vegetation areas from satellite imagery using digital landscape model and prior information, *International Archives of Photogrammetry and Remote Sensing*, 32, 167–174, 1999.

Isaaks, E.H. and Srivastava, R.M., *An Introduction to Applied Geostatistics*, Oxford: Oxford University Press, 1989.

Jacobson, C.E. and Dawson, M.R., Structural and metamorphic evolution of the Orocopia Schist and related rocks, southern California: Evidence for late movement on the Orocopia fault, *Tectonics*, 14, 933–944, 1995.

Journel, A.G. and Huijbregts, C.J., *Mining Geostatistics*, Cambridge, MA: Academic Press, 1981.

Maguire, D., *GIS, Spatial Analysis, and Modeling*, Redlands, CA: ESRI Press, 2005.

Noel, C., *Statistics for Spatial Data*, New York: Wiley, 1991.

Star, J. and Estes, J., *Geographic Information Systems: An Introduction*, Englewood Cliffs, NJ: Prentice-Hall, 1990.

Stein, M.L., *Statistical Interpolation of Spatial Data: Some Theory for Kriging*, New York: Springer, 1999.

Tian, B., Applications of GIS in structural geology: A case study on the structural geology of southern California and southwestern Arizona, *The 110th Annual Meeting of the Iowa Academy of Science*, Mason City, IA, April 25, 1998.

Tian, B., Computer mapping, data analysis, 3D modeling, and visualization of the geology in the Picacho-Trigo Mountains area of southern California and southwestern Arizona, M.S. Thesis, Iowa State University, IA, December 1998.

Tian, B., Enhancing understanding of environmental problems and finding solutions with GIS technology, *21st International Interdisciplinary Conference on the Environment*, San Juan, Puerto Rico, June 10–13, 2015.

Tian, B., Cudney, J., and Chulick, J., Collecting, analyzing, and visualizing environmental data using innovative GIS techniques at camp sibert, *International Journal of Environmental Science and Development*, 3(3), 211–216, 2012.

Tian, B. and Jacobson, C.E., Mapping, analysis, and 3D modeling of the geology in southern California with GIS, *10th Annual Graduate Student Seminar*, Iowa State University, March 28, 1998.

Tian, B. and Nimmer, P., Data analysis and visualization with GIS: Case studies on the geologic project in southeastern California and the environmental project in Brunswick, Maine, *The 15th NY State GIS Conference*, New York, October 1999.

Tian, B. and Nimmer, P., Integrating ArcView GIS with other applications to better manage and visualize environmental data, *ESRI International GIS User Conference*, June 2000.

Tian, B., Nimmer, P. and Easterday, A., 3D modeling and visualization of geologic and environmental data with GIS and GMS, *ESRI International GIS User Conference*, July 2001.

Tian, B. and Wu, H., Geochemistry of the two blueschist belts and implications for the Early Paleozoic tectonic evolution of the North Qilian Mts., China, in Wu, H., Tian, B., and Liu, Y. (eds.), *Very Low-Grade Metamorphism: Mechanisms and Geologic Applications, IGCP Project 294 International Symposium*, Seismological Press, pp. 92–116, March 1994.

Thomas, R.W. and Huggett, R.J., *Modeling in Geography: A Mathematical Approach*, London: Harper and Row, 1980.

Index

Printed and bound by CPI Group (UK) Ltd, Croydon, CR0 4YY

01/11/2024

01782619-0008